Reengineering the 2010 Census: Risks and Challenges

Panel on Research on Future Census Methods

Daniel L. Cork, Michael L. Cohen, and Benjamin F. King, *Editors*

Committee on National Statistics

Division of Behavioral and Social Sciences and Education

NATIONAL RESEARCH COUNCIL
OF THE NATIONAL ACADEMIES

THE NATIONAL ACADEMIES PRESS
Washington, D.C.
www.nap.edu

THE NATIONAL ACADEMIES PRESS 500 Fifth Street, NW Washington, DC 20001

NOTICE: The project that is the subject of this report was approved by the Governing Board of the National Research Council, whose members are drawn from the councils of the National Academy of Sciences, the National Academy of Engineering, and the Institute of Medicine. The members of the committee responsible for the report were chosen for their special competences and with regard for appropriate balance.

The project that is the subject of this report was supported by contract no. 50-YABC-8-66016 between the National Academy of Sciences and the U.S. Census Bureau. Any opinions, findings, conclusions, or recommendations expressed in this publication are those of the author(s) and do not necessarily reflect the views of the organizations or agencies that provided support for the project.

International Standard Book Number 0-309-09189-6 (Book)
International Standard Book Number 0-309-53146-2 (PDF)

Library of Congress Control Number 2004103813

Additional copies of this report are available from the National Academies Press, 500 Fifth Street, NW, Washington, D.C. 20001; (202) 334-3096; Internet, http://www.nap.edu.

Suggested citation: National Research Council (2004). *Reengineering the 2010 Census: Risks and Challenges.* Panel on Research on Future Census Methods. Daniel L. Cork, Michael L. Cohen, and Benjamin F. King, eds. Committee on National Statistics, Division of Behavioral and Social Sciences and Education. Washington, DC: The National Academies Press.

THE NATIONAL ACADEMIES

Advisers to the Nation on Science, Engineering, and Medicine

The **National Academy of Sciences** is a private, nonprofit, self-perpetuating society of distinguished scholars engaged in scientific and engineering research, dedicated to the furtherance of science and technology and to their use for the general welfare. Upon the authority of the charter granted to it by the Congress in 1863, the Academy has a mandate that requires it to advise the federal government on scientific and technical matters. Dr. Bruce M. Alberts is president of the National Academy of Sciences.

The **National Academy of Engineering** was established in 1964, under the charter of the National Academy of Sciences, as a parallel organization of outstanding engineers. It is autonomous in its administration and in the selection of its members, sharing with the National Academy of Sciences the responsibility for advising the federal government. The National Academy of Engineering also sponsors engineering programs aimed at meeting national needs, encourages education and research, and recognizes the superior achievements of engineers. Dr. Wm. A. Wulf is president of the National Academy of Engineering.

The **Institute of Medicine** was established in 1970 by the National Academy of Sciences to secure the services of eminent members of appropriate professions in the examination of policy matters pertaining to the health of the public. The Institute acts under the responsibility given to the National Academy of Sciences by its congressional charter to be an adviser to the federal government and, upon its own initiative, to identify issues of medical care, research, and education. Dr. Harvey V. Fineberg is president of the Institute of Medicine.

The **National Research Council** was organized by the National Academy of Sciences in 1916 to associate the broad community of science and technology with the Academy's purposes of furthering knowledge and advising the federal government. Functioning in accordance with general policies determined by the Academy, the Council has become the principal operating agency of both the National Academy of Sciences and the National Academy of Engineering in providing services to the government, the public, and the scientific and engineering communities. The Council is administered jointly by both Academies and the Institute of Medicine. Dr. Bruce M. Alberts and Dr. Wm. A. Wulf are chair and vice chair, respectively, of the National Research Council.

www.national-academies.org

PANEL ON RESEARCH ON FUTURE CENSUS METHODS

BENJAMIN F. KING *(Chair)*, Delray Beach, Florida
DAVID A. BINDER, Methodology Branch, Statistics Canada, Ottawa
MICK P. COUPER, Survey Research Center, University of Michigan, and Joint Program in Survey Methodology, University of Maryland
C.A. IRVINE, San Diego, California
WILLIAM KALSBEEK,* Department of Biostatistics, University of North Carolina, Chapel Hill
SALLIE KELLER-MCNULTY, Statistical Sciences Group, Los Alamos National Laboratories, Los Alamos, New Mexico
GEORGE T. LIGLER, Potomac, Maryland
MICHAEL M. MEYER, Intelligent Results, Inc., Seattle, Washington
DARYL PREGIBON,** Google, New York City
KEITH F. RUST, Westat, Inc., Rockville, Maryland
JOSEPH J. SALVO, Population Division, Department of City Planning, New York City
JOSEPH L. SCHAFER, Department of Statistics, Pennsylvania State University
ALLEN L. SCHIRM, Mathematica Policy Research, Inc., Washington, DC
JOSEPH H. SEDRANSK, Department of Statistics, Case Western Reserve University
C. MATTHEW SNIPP, Department of Sociology, Stanford University
DONALD YLVISAKER, Department of Statistics, University of California, Los Angeles
ALAN M. ZASLAVSKY, Department of Health Care Policy, Harvard Medical School

MICHAEL L. COHEN, *Study Director*
DANIEL L. CORK, *Study Director*
SHOREH ELHAMI, *Consultant*
AGNES E. GASKIN, *Senior Project Assistant*

*Served until March 2002
**Served until March 2001

COMMITTEE ON NATIONAL STATISTICS
2004

JOHN E. ROLPH *(Chair)*, Marshall School of Business, University of Southern California
JOSEPH G. ALTONJI, Department of Economics, Yale University
ROBERT M. BELL, AT&T Labs—Research, Florham Park, New Jersey
LAWRENCE D. BROWN, Department of Statistics, The Wharton School, University of Pennsylvania
ROBERT M. GROVES, Survey Research Center, University of Michigan, and Joint Program in Survey Methodology, University of Maryland
JOHN C. HALTIWANGER, Department of Economics, University of Maryland
PAUL W. HOLLAND, Educational Testing Service, Princeton, New Jersey
JOEL L. HOROWITZ, Department of Economics, Northwestern University
WILLIAM KALSBEEK, Department of Biostatistics, University of North Carolina, Chapel Hill
ARLEEN LEIBOWITZ, School of Public Policy and Social Research, University of California, Los Angeles
VIJAYAN NAIR, Department of Statistics and Department of Industrial and Operations Engineering, University of Michigan
DARYL PREGIBON, Google, New York City
KENNETH PREWITT, School of Public Affairs, Columbia University
NORA CATE SCHAEFFER, Department of Sociology, University of Wisconsin, Madison

JAMES F. HINCHMAN, *Acting Director*
CONSTANCE F. CITRO, *Acting Chief of Staff*

Acknowledgments

THE PANEL ON RESEARCH ON FUTURE CENSUS METHODS of the Committee on National Statistics is pleased to submit this final report and wishes to thank the many people who have contributed to our work over the panel's lifetime.

We thank the staff of the U.S. Census Bureau, under the leadership of director C. Louis Kincannon, deputy director Hermann Habermann, former director Kenneth Prewitt, and former acting director William Barron, for their interactions with the panel. In particular, we appreciate the efforts of Preston Jay Waite, associate director for decennial census. Rajendra Singh, the panel's lead liaison with the Census Bureau, and Philip Gbur provided useful assistance. In plenary sessions and in smaller working group activities, the panel has also benefited from its interaction with other talented members of the Census Bureau staff, including Teresa Angueira, Andrea Brinson, Jon Clark, Dave Galdi, Nancy Gordon, Edison Gore, Joan Hill, Howard Hogan, Arnold Jackson, Dean Judson, Ruth Ann Killion, Joe Knott, Donna Kostanich, Juanita Lott, Robert Marx, Fay Nash, Alfredo Navarro, Sally Obenski, Ed Pike, Linda Pike, Jim Treat, Alan Tupek, Carol Van Horn, Frank Vitrano, and Tracy Wessler.

The tragic death of Charles H. "Chip" Alexander, Jr., in early September 2002 was an incalculable loss for the entire research community surrounding the decennial census and its related programs. The chief statistical methodologist for the American Community Survey (ACS), Chip was also the panel's designated liaison on ACS matters. One true pleasure of service on this panel was the opportunity for interaction with someone of Chip's great knowledge and good humor, and we join his friends and colleagues in mourning his loss.

Our panel colleague Joseph Salvo, of the New York City Department of City Planning, ably chaired a working group to evaluate the Local Update of Census Addresses (LUCA) program from the local government perspective. Jointly sponsored by this panel and our sister panel, the Panel to Review the 2000 Census, LUCA working group members drew from their first-hand expertise in documenting their LUCA experience in case study form. We thank the members of this group—Shoreh Elhami, Abby Hughes, Terry Jackson, Tim Koss, and Harry Wolfe—and working group consultant Patricia Becker for their efforts, a solid reference work for our panel and the entire research community.

In particular, we have benefited greatly from the continuing consultation of LUCA working group member Shoreh Elhami, of the Delaware County (Ohio) Auditor's Office. A current member of the National Research Council's Mapping Science Committee, her expertise on census and geography matters from the local government perspective has enriched our discussions of the Census Bureau's plans to modernize their geographic resources.

In April 2001 the panel opened its first examination of the proposed MAF/TIGER Enhancements Program by inviting a distinguished set of discussants to share their opinions on the proposed plans. In addition to Shoreh Elhami, this roster of discussants included Rick Ayers (Environmental Systems Research Institute, Inc.), Donald Cooke (Geographic Data Technology, Inc.), Michel Lettre (State of Maryland), and Sarah Nusser (Department of Statistics, Iowa State University). We thank them for their time and their talents.

At the request of the Census Bureau, panel staff organized a meeting on September 10, 2003, dealing specifically with the Census Bureau's plans to redesign the database structure for its geographic resources (Master Address File and TIGER geographic database). Conducted by the Census Bureau, the meeting supplemented expertise on the panel with additional experts in computer science, software engineering, and geography. Panel members Al Irvine and Mike Meyer participated in the meeting; we thank invitees Michael Goodchild (University of California, Santa Barbara), Les Miller (Iowa State University and American Statistical Association/National Science Foundation Census Fellow), Jesse Poore (University of Tennessee), and Allan Wilks (AT&T Laboratories–Research) for their participation and discussion.

We are grateful to our colleagues on the companion Panel to Review the 2000 Census and to its chair, Janet Norwood, for their assistance and contributions over the course of the panel's study. Members of our panel joined members and staff of the Norwood panel to visit local and regional census offices during the 2000 census. Since those early days, both panels have been continually updated on each other's progress. In particular, the Norwood panel's detailed discussion of the 2000 Accuracy and Coverage Evaluation and the possible statistical adjustment of the 2000 census figures has been most helpful to us in suggesting priorities for coverage measurement in 2010.

Over the years, the panel has greatly benefited from good relations and insightful interaction with the broader census community. Terri Ann Lowenthal's "Census 2000 News Briefs" have been a most helpful resource and an important communications channel, and we appreciate her efforts. We have learned much from our discussions with relevant staff of the U.S. House Subcommittee on the Census of the Committee on Government Reform (and its successor subcommittees with census oversight authority), particularly David McMillen and former subcommittee staffer Michael Miguel. We have appreciated our interaction with census-related staff of the U.S. General Accounting Office, including Robert Parker and Ty Mitchell. We also thank Ed Spar of the Council of Professional Associations on Federal Statistics and Susan Schechter and Katherine Wallman of the U.S. Office of Management and Budget for their participation in panel meetings.

Logistical arrangements for panel activities were made with great skill by Agnes Gaskin, senior project assistant. Research assistant Marisa Gerstein deserves thanks for her help with maintaining an archive of materials related to both this panel and the Panel to Review the 2000 Census. Former CNSTAT staff member Carrie Muntean, now stationed with the U.S. Foreign Service, did exemplary work for both panels and, in particular, with the LUCA working group. CNSTAT consultant Meyer Zitter's enthusiasm in collecting information for both panels is greatly appreciated. Cameron Fletcher, associate editor, and Christine McShane, senior editor of the reports office of the Division of Behavioral and Social Sciences and Education, contributed to this report with fine technical editing. Finally, virtually all of the activities of the Committee on National Statistics benefit greatly from the involvement and advice of Constance Citro, senior program officer. Given her role as study director of the Panel to Review the 2000 Census, we have drawn quite heavily on her wisdom and benefited from her contributions.

Finally, I represent the whole panel in expressing our gratitude to Daniel Cork and to Michael Cohen, the codirectors of this study, for their invaluable assistance in all aspects of our work. Neither this report nor the interim and letter reports that have preceded it would have been possible without their excellent liaison activities with the Census Bureau, their able handling of the logistics of our meetings, their up-to-date reporting to distant panel members of all developments in the Census Bureau's planning for 2010, and their translation into readable prose of our reactions to and recommendations for the process as it has unfolded. Personally, it has been a great pleasure to work with them.

This report has been reviewed in draft form by individuals chosen for their diverse perspectives and technical expertise, in accordance with procedures approved by the Report Review Committee of the National Research Council (NRC). The purpose of this independent review is to provide candid and critical comments that will assist the institution in making the published report as sound as possible and to ensure that the report meets institutional standards for objectivity, evidence, and responsiveness to the study charge. The review comments and draft manuscript remain confidential to protect the integrity of the deliberative process.

We thank the following individuals for their participation in the review of this report: Barbara A. Bailar, statistical consultant, Washington, DC; Barbara Everitt Bryant, University of

Michigan Business School; Don A. Dillman, Departments of Sociology and Community Rural Sociology and Social Economic Sciences Research Center, Washington State University; Michael Hout, Department of Sociology, University of California, Berkeley; Janet Norwood, consultant, Chevy Chase, MD; and Halliman H. Winsborough, Department of Sociology and Center for Demography and Ecology, University of Wisconsin–Madison.

Although the reviewers listed above provided many constructive comments and suggestions, they were not asked to endorse the conclusions or recommendations nor did they see the final draft of the report before its release. The review of the report was overseen by Robert Hauser, Center for Demography, University of Wisconsin–Madison. Appointed by the National Research Council, he was responsible for making certain that an independent examination of the report was carried out in accordance with institutional procedures and that all review comments were carefully considered. Responsibility for the final content of this report rests entirely with the authoring panel and the institution.

Benjamin F. King, *Chair*
Panel on Research on Future Census Methods

Contents

Acronyms and Abbreviations

ACE	Accuracy and Coverage Evaluation
ACF	Address Control File
ACS	American Community Survey
ALMI	Automated Listing and Mapping Instrument
AREX 2000	Administrative Records Experiment (2000)
BAS	Boundary and Annexation Survey
BSA	basic street address
C2SS	Census 2000 Supplementary Survey
CAI	computer-assisted interviewing
CAPI	computer-assisted personal interviewing
CATI	computer-assisted telephone interviewing
CAUS	Community Address Updating System
CEFU	coverage edit follow-up
CIFU	coverage improvement follow-up
CIO	chief information officer
CMM	Capability Maturity Model
CNSTAT	Committee on National Statistics
COTS	commercial off-the-shelf
CPS	Current Population Survey
DADS	Data Access and Dissemination System
DCS 2000	Data Capture System 2000
DEX	Digital Exchange
DMAF	Decennial Master Address File

DSE	dual-systems estimation
DSF	Delivery Sequence File
ESCAP	Executive Steering Committee on ACE Policy
FEAF	Federal Enterprise Architecture Framework
FIPS	Federal Information Processing Standards
GAO	U.S. General Accounting Office
GBF/DIME	Geographic Base File/Dual Independent Map Encoding
GIS	geographic information systems
GPS	global positioning system
GQ	group quarters
GSS	Geographic Support System
ICM	Integrated Coverage Measurement
IDC/IQA	Internet Data Collection/Internet Questionnaire Assistance
IDEF0	Integration Definition for Function Modeling
IRS	Internal Revenue Service
IT	information technology
IVR	interactive voice response
LCO	local census office
LUCA	Local Update of Census Addresses
MAF	Master Address File
MAF/TIGER	Master Address File/Topologically Integrated Geographic Encoding and Referencing System
MAFGOR	MAF Geocoding Office Resolution
MaRCS	Matching and Review Coding System
MCD	minor civil division; mobile computing device (Census Bureau usage)
MIS 2000	Management Information System 2000
MTAIP	MAF/TIGER Accuracy Improvement Project
MTEP	MAF/TIGER Enhancements Program
NCT	National Census Test (2003)
NRC	National Research Council
NRFU	nonresponse follow-up
OCR	optical character recognition
OCS 2000	Operations Control System 2000
OIG	Office of Inspector General (U.S. Department of Commerce)

OMB	Office of Management and Budget
OMR	optical mark recognition
PALS	Program for Address List Supplementation
PAMS/ADAMS	Pre-Appointment Management System/Automated Decennial Administrative Management System
PCD	portable computing device
PDA	personal digital assistant
PES	postenumeration survey
PRED	Planning, Research, and Evaluation Division (U.S. Census Bureau)
PSA	primary selection algorithm
RFP	request for proposals
RMIE	Response Mode and Incentive Experiment
SIPP	Survey of Income and Program Participation
SNRFU	sampling for nonresponse follow-up
StARS	Statistical Administrative Records System
TEA	type of enumeration area
TIGER	Topologically Integrated Geographic Encoding and Referencing System
TMU	Targeted Map Update
TQA/CEFU	Telephone Questionnaire Assistance/Coverage Edit Follow-Up
USGS	United States Geological Survey
USPS	United States Postal Service

Executive Summary

At the request of the U.S. Census Bureau, the Panel on Research on Future Census Methods was organized to review the early planning process for the 2010 census. Its work includes observing the operation of the 2000 census, deriving lessons for 2010, and advising on effective evaluations and tests. The panel has previously issued two interim reports (National Research Council, 2000a, 2003a) and a letter report (National Research Council, 2001c), and this is our final report.

EMERGING STRUCTURE OF THE 2010 CENSUS

The Census Bureau's current plans for the 2010 census are predicated on the completion of three major initiatives, which the Bureau has described as a "three-legged stool":

1. *MAF/TIGER Enhancements Program.* A specific set of improvements has been proposed to the Census Bureau's address list (Master Address File, or MAF) and geographic database (Topologically Integrated Geographic Encoding and Referencing System, or TIGER).
2. *American Community Survey (ACS).* The decennial census long form will be replaced by a continuous survey, thus permitting a short-form-only census in 2010. The ACS covers the same social, economic, and demographic data as

the census long form but will provide estimates in a more timely manner.

3. *Early Integrated Planning.* The Census Bureau hopes that early attention to planning will make census tests leading up to 2010 more informative and useful.

The Census Bureau's emerging 2010 census plan also includes the development of portable computing devices (PCDs) for use in nonresponse follow-up work and the increased use of multiple response modes (mail, Internet, and telephone).

CENSUS REENGINEERING: A PROCESS AT RISK

The Census Bureau has advanced an ambitious vision for the 2010 decennial census and—as our previous reports and the balance of this report suggest—the panel strongly supports the major aims of the plan. The implementation of the ACS and, with it, the separation of the long form from the census process are very good concepts; the Bureau's address and geographic databases are in dire need of comprehensive update; and the implementation of new technologies in census-taking is crucial to improving the accuracy of the count. There is much to like about the emerging plans for the 2010 census, and we strongly support these efforts toward a modernized and improved census in 2010. To this end, the Census Bureau's focus on planning early in the decennial cycle is highly commendable.

However, based on the information available, the panel finds that the reengineering of the 2010 census is a process at high risk. The major initiatives of the 2010 census plan—the MAF/TIGER Enhancements Program and the American Community Survey—are intended to reduce operational risk in the census in the long term. However, their implementation in the short term necessarily creates unique risks and challenges. In addition, adoption of new technology is inherently risky, particularly when done on the tight schedule and large scale of the decennial census.

To be clear, our conclusion that the reengineering of the 2010 census is a process at risk should not be interpreted as a conclusion that the 2010 census is irrevocably headed for serious problems. It is not an argument that Census Bureau resources are being focused on plans that are wrong and, therefore, that spending should be trimmed. Quite to the contrary, our argument is meant to underscore the importance of a rigorous and amply funded planning and testing cycle for the 2010 census.

The reengineering of the 2010 census faces two paramount risks that are, to a large extent, out of the control of the Census Bureau. Those risks are, first, that funding will not be available for major components of the census plan or will be available at unpredictable levels and, second, that the decision on the final design of the census (particularly the role of the ACS or a census long form) will not be finalized or will shift with time. Decisions on funding and overall design direction are ultimately up to Congress and the administration. However, though the Census Bureau may not have control over funding decisions, the panel believes that the Bureau must take a more active role in informing the funding and decision process in at least two ways.

First, as we emphasize throughout this report, the Bureau must develop a sound research and evidentiary base for its 2010 census plan, thus making a stronger and more compelling case for sustained long-term funding. Building this research base should include carefully examining operational data from the 2000 census to guide planned practice for 2010 and fully exploring the potential of new tools for evaluation, among them the Master Trace Sample containing results of all census operations for a limited national subset. Much work also remains in integrating and mapping the logical and technical infrastructures of the entire census process, and in developing a rigorous and timely testing and evaluation program for new census systems and techniques. The consequences of failing to develop a strong research base for the 2010 census are significant: repeating past census processes that may be inefficient or suboptimal, conducting a census with methods that are out of step with the dynamics of the population it is intended to count, making limited technological innovations that may not match real needs, and marking a flawed beginning for the 2020 census.

Second, the Bureau should be explicit in identifying, articulating, and quantifying the consequences associated with risks in the census process—for instance, the impact of reduced funding on the quality of ACS estimates for small geographic areas and population groups. Failure to reach consensus on the role of the ACS in the census process raises the undesirable prospect of a reversion to the long form, possibly late in the census process and therefore implemented in a rushed manner, thus incurring the same nonresponse and data quality problems as were experienced with the 2000 long form. Such failure would impair other parts of the census plan, including effective use of PCDs. More significantly, failure to reach closure on census design leaves open the possibility that the detailed socioeconomic and demographic characteristics measured by the current census long form may not be estimated at all in 2010, an unacceptable outcome for many reasons.

SPECIFIC RISK CATEGORIES AND MITIGATION STRATEGIES

Beyond the broad risks of funding and design selection, the 2010 census planning process faces many risks of a more specific nature. Some of them are acknowledged in the Bureau's draft risk management plan, but many are not. Based on the information known to us, we find that the 2010 census reengineering process may be jeopardized in the following areas, among others.

Modernizing Geographic Resources

The panel believes that the process by which the Master Address File is updated and improved is severely at risk. The Census Bureau's current approach relies principally on updates from U.S. Postal Service files, effectively treating MAF updating as "routine maintenance." Moreover, the Bureau appears set to rely on a complete block canvass of mailing addresses, a costly operation just before the census. Absent a strong focus on enhancing the MAF in its own right, throughout the decade and independent of presumed benefits from linkage to a realigned TIGER database, the 2010 census may be conducted with an address source that has unacceptable levels of housing unit duplication in some areas and coverage gaps in others.

In the panel's assessment, it is particularly critical that the Census Bureau develop a comprehensive plan for updating and improving the Master Address File (Recommendation 3.1). A centralized staff position to coordinate housing unit definition and listing (Recommendation 3.2) would help create a quality address list for 2010. The panel also suggests research and analysis of various possible sources of address updates: work with the Postal Service on assessing the quality of the Delivery Sequence File (Recommendation 3.3), analysis of the possible contribution of the Community Address Updating System (3.4), and justification of the Bureau's plan to implement a complete block canvass (3.6). Further analysis of MAF data from the 2000 census is a crucial learning tool (3.7).

The Bureau's current MAF/TIGER Enhancements Program focuses on the realignment of TIGER features and modernization of the TIGER database structure; each of these tasks has considerable associated risk. First, the initial realignment of TIGER geographic features to be consistent with GPS coordinates may not be completed in time, or change detection for new features after the initial realignment may not be properly performed. These outcomes would have a negative impact on plans for the use of personal computing devices by field enumerators and would lead to continued errors in the geocoding of addresses in the census and in nonresponse follow-up operations. Second, the conversion of the MAF/TIGER database from its current homegrown format to a modern, object-oriented computing environment may be slower or more difficult than anticipated. The transition will be more risky if the Census Bureau attempts the conversion en masse, rather than via a more carefully designed software reengineering process with ample testing. Careful planning will be essential to keeping TIGER modernization on track. We also suggest that the development of MAF/TIGER support software could be an opportunity to build stronger ties with software developers outside the Bureau (Recommendation 6.3).

The Census Bureau's draft risk management plan discusses the possibility of alienating key stakeholders, including local and tribal governments. Alienation of local authorities is a risk, to be certain, but a more fundamental risk is failure to fully involve them in census design and operations. We urge the Census Bureau to develop a complete plan for its partnerships with local and tribal governments, with particular regard to address list updating (Recommendation 3.5) and more generally. In our assessment, cultivation of strong local and tribal partnerships will also help redefine enumeration techniques for group quarters and other special populations, tailor enumeration techniques to specific areas within localities, and foster the acceptance and use of the ACS.

American Community Survey

As discussed previously, the introduction of the ACS and elimination of the census long form are the most fundamental factors in determining the final design of the 2010 census, and delay in finalizing that design is one of the most fundamental risks. Accordingly, the panel emphasizes the need for a clear and early decision (Recommendation 4.4) and for contingency plans for the collection of traditional long-form data should full ACS funding not be forthcoming (4.5).

The panel believes that development of a strong research and evaluation program for the ACS is important in several respects. Resolution of issues regarding estimation techniques based on a continuous survey like the ACS and further exploration of the relationship between the ACS and other federal surveys are essential to winning support for the ACS and to its adoption by data users. Lack of a full ACS research agenda may also pose longer-term risks to the quality and usefulness of the survey, hindering potential ties between the ACS and programs for producing postcensal population and demographic analysis estimates. The panel recommends continued research on the relative quality of ACS and census long-form-sample estimates (Recommendation 4.1), development of a "user's guide" to ACS data (4.3), and sharing of detailed ACS data with local data analysts and the broader research community (4.2).

Enumeration and Data-Processing Methods

The Census Bureau's plans for the use of portable computing devices (PCDs) in the 2010 census are a particularly exciting part of a reengineered census, but the plans also entail risks associated with the implementation of new technologies. Perhaps most significant is the risk that the Bureau may fail to fully understand the direction in which the technology is moving and thus may spend its resources testing devices that are inferior to those that will be available in 2010, in terms of both size and computing capacity. A consequence of this error is that wrong and misleading conclusions would be drawn about the real potential for portable computing devices to improve census data collection. Accordingly, we recommend that the Bureau conduct a rigorous test of PCDs for data collection, including tests using current high-end devices that may be closer to what will be widely available at time of procurement (Recommendation 5.1).

A second risk inherent with the PCD technology lies in making the decision to purchase too early and without fully specified requirements, resulting in the possible selection of obsolete or inadequate devices. Third, and related, is the risk that the Census Bureau may not use the sheer size of its order (perhaps on the order of 500,000 devices) to obtain devices tailored to census needs, as opposed to buying only what is commercially available off the shelf. Finally, given the fact that the principal users of the devices will be the large corps of temporary enumerators (with limited training), there is the risk that Census Bureau PCD development will not take human factors into sufficient consideration. These risks are significant, and there are no set guidelines we can offer regarding the optimal time to buy. In our assessment, the best way to mitigate these risks is to focus on the detailed specifications for the devices—defining exactly what the devices must be able to do—and try to tailor the final devices to those specifications (and not vice versa; Recommendation 5.2). Particular attention must also be paid to designing a complete testing protocol for PCD software and hardware components (Recommendation 5.3).

It is a basic truth that some people, households, and areas are inherently more difficult to count in the census than others. The experience of the 2000 census suggests several risks for 2010 planning related to hard-to-count populations. Significant among these is the population living in group quarters (places such as hospitals, dormitories, and prisons). In 2000, as in previous censuses, procedures for enumerating group quarters were conducted separate from the rest of the census process. Group quarters listings were not reconciled with the MAF for households, and little effort was given to the challenges of enumerating different types of group quarters. The enumeration processes for group quarters were not well controlled. Continuing with this approach incurs the risk of duplication and other enumeration errors and ineffective coverage of this small but important population group. We strongly recommend comprehensive reexamination of the definition, listing, and enumeration procedures for special places and group quarters (Recommendation 5.4).

Group quarters are not the only populations that have traditionally posed difficulties; others include immigrant communities, irregular multiunit housing structures, gated communities, colonias along the U.S.-Mexico border, and the homeless. Enumeration efforts for these populations may be compromised by failure to clarify the definition and presentation to respondents of the residence rules for the decennial census. Consequences associated with such a failure include poor-quality data, failure to meet consumer needs, and continued differential undercount. There is no definitive advice we can offer about the best way to count these groups; what we do suggest is that dialogue and plans for counting them begin early in the census planning cycle rather than being saved as last-minute considerations (Recommendation 5.5).

Though the 2010 census is still expected to be conducted largely by mail, the Census Bureau will likely promote other modes by which respondents can return their forms and also introduce different contact strategies to reach respondents. In particular, use of the Internet to reply to the short-form-only census will likely be encouraged; interactive voice response through an automated telephone system may also be used, though that technology has experienced difficulty in early testing. Additional response modes and other programs such as a repeat of the 2000 "Be Counted" campaign (by which people who believed they were missed in the census could pick up a form in public locations) may increase public cooperation in the census. However, they also raise the risk of higher levels of duplication, of both persons and housing units. The Census Bureau had to implement ad hoc unduplication techniques as the 2000 census was processed, and some of these activities may be formalized in 2010. The panel notes the need for a research plan for unduplication techniques and the need to test proposed techniques in the 2006 census test (Recommendation 5.7). One respondent contact strategy of particular merit is the sending of replacement questionnaires to nonresponding households; plans to do this in 2000 had to be abandoned when it became apparent that it could not be done in a timely fashion. The Census Bureau must proceed quickly to find ways to effectively operationalize a targeted replacement questionnaire in 2010 (Recommendation 5.6).

For the 2000 census, the Census Bureau chose to complete nonresponse follow-up activities very quickly and rely on long-standing imputation techniques to fill in missing questionnaire items and, in some cases, to impute household size when no information was available for a presumed-occupied unit. These techniques came under scrutiny following the census when the state of Utah challenged the inclusion of some types of imputations in apportionment totals. Although the U.S. Supreme Court ultimately upheld the use of existing Bureau imputation practices, the debate suggests the need to revisit the techniques, including the "hot-deck" methodology that has been used by the Census Bureau for several decades. Specifically, the panel strongly urges the Census Bureau to investigate further the costs and benefits of the basic trade-off between continuing field nonresponse follow-up work versus imputation for nonresponse (Recommendation 5.8) and to further study the effect of imputation techniques on the distribution of census data items (5.9).

Technical Infrastructure

Conduct of the decennial census requires a sound technical infrastructure—the amalgam of people, computer hardware, software programs, and telecommunication networks that facilitate the flow and processing of information from beginning to end. In 2000, many of the systems that the Census Bureau used were ultimately successful but were developed at great risk, often hastily and without opportunity for full testing. The Census Bureau has begun efforts toward modeling the logical infrastructure—the information blueprint that diagrams all the informational dependencies between pieces of the census process—and using that logical infrastructure to guide the development of the physical technical architecture. The panel strongly endorses these efforts (Recommendation 6.1) and notes the need for strong institutional commitment and "championship" of architecture redesign. To this end, the panel advocates the creation of the position of system architect of the decennial census to coordinate this effort, and further recommends that subsystem architects for MAF/TIGER and field operations (PCDs) be recruited (Recommendation 6.2). The Census Bureau's draft architecture documents suggest that the Bureau's efforts have not yet reached the stage at which real reengineering can take place, but this will hopefully be resolved quickly with further experience with the modeling techniques. Failure to achieve the full potential of architecture modeling would incur severe risks: systems may be ill-suited to handle 2010 census process needs, may fail during actual census operations due to lack of proper testing, and may not properly interoperate with each other.

Generally, the Census Bureau has expressed the desire to improve its capabilities in software engineering, motivated in particular by the need to redesign the database structure underlying the MAF/TIGER system from a home-grown environment to one based on commercial products. While noting that improving software engineering practices is difficult in its own right, much less on the tight schedule and amid the other demands of the decennial census, the panel supports the Bureau's effort to improve its software standards (Recommendation 6.4) and, in particular, urges greater attention to the Bureau's protocols for computer hardware and software testing (6.5).

Coverage Measurement

Disputes over the role of sampling methods and the use of dual-systems estimation (based on matching an independent postenumeration survey to census returns) were the dominant force in planning the 2000 census. From all indications, it is the role of the American Community Survey and the prospective replacement of the census long form that will be the major force in deciding the overall shape of the 2010 census, rather than considerations of coverage measurement and evaluation. After the statistical adjustment battles preceding and following the 2000 census, the Census Bureau may be understandably reluctant to take up active debate on coverage techniques for 2010. However, this reluctance incurs the risk that a comprehensive plan for the measurement and assessment of census coverage in 2010 will be deferred until late in the census process. It is essential that the Census Bureau have the means to determine the accuracy of its count, and a late-course fallback to the same Accuracy and Coverage Evaluation methodology used in 2000 could be unfortunate, particularly if research on problems raised by the 2000 census experience are not addressed in the intervening decade.

In the panel's assessment, the coverage measurement program in 2010 need not take the same exact shape as that of 2000; what is important is that plans for the program are developed early, and that techniques are tested in 2006 and in the 2008 dress rehearsal. The Panel to Review the 2000 Census has comprehensively reviewed the 2000 Accuracy and Coverage Evaluation research and suggested changes and improvements; these should be implemented, to the extent that a postenumeration survey is part of the 2010 coverage plan. The panel encourages further research on the data and assumptions that support demographic analysis estimates (Recommendation 7.1), which have served as an important coverage benchmark in recent censuses. The panel also encourages further work on methods based on administrative records; whether or not there is a role for such methods in the conduct of the 2010 census, administrative records work should at least be a major experiment in the 2010 census as it was in 2000.

General Research, Evaluation, and Testing

As evidenced by the common theme in many of our recommendations, the panel believes that research and evaluation are essential not only to the diagnosis of risks inherent in the census process but also to their mitigation. Accordingly, the Census Bureau should materially strengthen and extend its program of evaluations (Recommendation 8.1). Evaluation should play a central role in operations rather than being relegated to a peripheral, post hoc role. In the past, the Census Bureau's planning and research entities have operated at either a high level of focus (e.g, articulation of broad objectives such as the "three-legged stool" components without laying out a clear base in empirical evidence) or at a microlevel that tends toward detailed accounting of individual census processes. As it designs its research and evaluation program, the Census Bureau should work to bridge the gap between research and operations in the census process; evaluations should be forward-looking and designed to inform and satisfy specific planning objectives.

The panel strongly encourages the further mining and reanalysis of operational data from the 2000 census to build a strong base for 2010 census planning (Recommendation 8.3). In particular, the panel urges the Bureau to make use of the Master Trace Sample, a compilation of data from many census operations, for a sample of the population that the panel has advocated in its previous reports (Recommendation 8.4). In addition to expanding the sample's scope to include key data such as results from the Accuracy and Coverage Evaluation (Recommendation 8.5), the panel strongly urges the Bureau to reconsider its current decision to limit access to the Master Trace Sample to internal Bureau users. Instead, the sample—with appropriate safeguards on confidentiality—should be accessible to the broader research community (Recommendation 8.6). As it designs its technical infrastructure and, hopefully, makes research a strong central focus, the Bureau should have as a long-term objective the maintenance of a Master Trace *System*—through which real-time evaluation could inform census operations even as the census is being fielded—rather than merely a *Sample* for which data are assembled after operations are completed (Recommendation 8.7).

A major focus of the Bureau's ongoing research and evaluation program should be the development of targeted methods for address list development and enumeration (Recommendation 8.2). Examples of these methods include targeting block canvass to verify address list entries to particular (e.g., high-growth) areas and expansion of update/leave enumeration (where a census enumerator drops questionnaires at housing units, which are then expected to mail them back) in areas where mail delivery may not be effective. Failure to implement these methods appropriately may result in increased costs and continued problems of enumeration in high-density areas with structures containing multiple (and not well-listed or easily differentiated) housing units.

The 2010 census is more imminent than many lay observers might expect; in particular, the number of opportunities for major census tests between now and 2010 is very limited. The 2004 test is currently being conducted, leaving only the 2006 census test and 2008 dress rehearsal as the major anticipated testing opportunities. The panel strongly encourages the Census Bureau to pursue smaller-scale testing as resources and timing permit, with the argument that not all census tests need to be part of a general, omnibus test census that is commonly the shape of the Bureau's major test opportunities. The Census Bureau has taken as a major goal the performance of a true dress rehearsal in 2008, in comparison to the 1998 dress rehearsal which was fundamentally an experimental test of competing census designs. The panel believes that this goal makes the 2006 census test much more important and crucial to a successful 2010 census. The 2006 census test should therefore be cast as a proof of concept, not a technical test; it must provide the basis for answering any remaining experimental questions in order to make the 2008 test a truly preoperational rehearsal, and it must be funded accordingly.

Part I

Background and General Planning

CHAPTER 1

The Panel on Research on Future Census Methods

AS THE 2000 CENSUS APPROACHED, the U.S. Census Bureau requested that the Committee on National Statistics (CNSTAT) of the National Research Council convene two panels, one to provide an independent and comprehensive review of the 2000 census and one to examine census conduct in 2000 with an eye toward the planning of the 2010 census. The Panel to Review the 2000 Census began work in 1998 (National Research Council, 2001a, 2004). Our Panel on Research on Future Census Methods began its operations in 1999 to assist the early planning efforts for the 2010 census, and this is our final report.

1–A CHARGE AND OPERATIONS OF THE PANEL

The Panel on Research on Future Census Methods has as its charge the following:

> The panel will review the plans for acquisition, analysis, and evaluation of research data needed to begin planning for the 2010 decennial during the 2000 census. The panel will suggest improvements and preferred approaches, observe the implementation of the 2000 census, suggest pri-

orities for analyzing the census experimental and tracking data, examine census accuracy, evaluate the research program results, and determine appropriate lessons for the 2010 census.

During the course of its work, the panel's charge evolved into a broad charter to review and advise on the emerging general plans for the 2010 census. This alteration in focus was enacted with the knowledge and consultation of the Census Bureau and through identification of a broader set of tasks in contract modifications with the Bureau. It also evolved naturally due to three factors.

First, this panel was formed very early relative to the census it examines, a factor that makes it unique in the experience of National Research Council panels regarding the decennial census. Our early start—beginning before the 2000 census—provided valuable opportunities to observe the census process, but also created unique challenges. The panel underwent a nearly yearlong hiatus in 2000, respecting the heavy demand on the Census Bureau and its senior staff during the active follow-up and processing of the 2000 census. In the summer and fall of 2001, the panel observed another period of relative dormancy at the Census Bureau's request, as the Bureau's executive and evaluation staff were committed to intensive research and deliberation over the question of statistical adjustment of 2000 census data for estimated undercount. The intensity of census coverage evaluation research in 2001 also delayed the Bureau's general program of operational and procedural evaluations of the 2000 census, which was to be the focus of the panel's activities under its original charge.

Second, the Census Bureau made an early start on active planning for the 2010 census. At the end of 2000, the Bureau sought the panel's assessment of the emerging "three-legged stool" plan for 2010 (consisting of the new American Community Survey, modernized geographic resources, and early planning and testing, and discussed further in Chapter 2). Since that time, the developing 2010 census plan and its major initiatives have been the primary areas of concentration in the panel's work. In cooperation with the Bureau, full-panel meetings during 2001 and 2002 were largely replaced by meetings of five working groups of panel members: Address List Development, American Community Survey, Enumeration Methods, Technical Infrastructure, and Coverage Evaluation and Statistical Infrastructure. Each of these working groups dealt with aspects of the developing 2010 census plan, and each was assigned liaison staff from the Census Bureau. The broad classification of topics across the working groups provides the basic structure for this final report.

Finally, the role and charge of our panel was influenced by the simultaneous operation and coordination of our panel and the Panel to Review the 2000 Census. The two panels were regularly apprised of each other's work during their tenures; members of our panel participated in some of the activities and workshop meetings of the 2000 census panel and vice versa. By their nature, the charges of the panels overlap, with both panels having at their core a mission to review the 2000 census and advise on possible changes. Given the presence of a standing panel to comprehensively review the 2000 census, it was natural for our panel to be (as we have colloquially been known since our founding) "the 2010 panel" and to take as a primary focus the developing plans and initiatives of the 2010 census.

In support of our charge, the panel conducted a variety of activities. The panel met in plenary session nine times during its tenure, but more often in small-group settings. Members of our panel joined the Panel to Review the 2000 Census to visit data capture centers, regional census offices, and local census offices during the conduct of the 2000 census. The two census panels held one joint meeting in March 2003 to hear final results from the Census Bureau's Accuracy and Coverage Evaluation Program and the Bureau's decision on statistical adjustment of census data for use in generating postcensal population estimates.

In addition, the two census panels jointly established a working group to evaluate the Local Update of Census Addresses (LUCA) Program, in which state, local, and tribal governments were able to review address lists or block-level population counts for their areas and suggest revisions. The Working Group on LUCA completed its report to both panels in early 2001 (Working Group on LUCA, 2001).

1–B PREVIOUS REPORTS OF THE PANEL

In February 2000, the panel issued its first interim report, *Designing the 2010 Census* (National Research Council, 2000a), based on early information gleaned from the panel's first two meetings. The report focused on priorities for the evaluation program for the 2000 census and argued in particular for the creation of a Master Trace Sample, collating information from many census operational databases for a sample of addresses.

In December 2000, the panel heard the Census Bureau's first presentation of its preliminary 2010 census strategy (the "three-legged stool" approach described in Section 2–C), and offered early feedback on the general strategy in a letter report to acting census director William Barron in February 2001 (National Research Council, 2001c).

The panel's second interim report, *Planning the 2010 Census*, was released in July 2003 (National Research Council, 2003a). In that report, the panel focused on three substantive areas: development of a technical infrastructure, modernization of geographic resources, and implementation of the American Community Survey (ACS). The report also provided brief comments on the plans for the 2003 and 2004 census tests.

1–C OVERVIEW OF THIS REPORT

This final report of the Panel on Research on Future Census Methods builds from and extends the material from the two interim reports that preceded it. Indeed, the text and structuring of some portions of the report are adapted directly from those reports, with appropriate revisions; we have chosen this approach (rather than incorporating the material through citation) in order to make this report as self-contained a document on 2010 census planning as possible.

The report is structured in three parts. Part I provides background and an overview of the Census Bureau's general plan for the 2010 census. Following the present chapter's synopsis of the panel's charge and activities, Chapter 2 describes the basic 2010 census strategy that has been developed by the Bureau. More significantly, Chapter 2 summarizes conclusions about the risks inherent in the 2010 census plan.

In Part II, we turn to specific issues in the design of the 2010 census as the plans have been developed to date. Chapter 3 discusses the Census Bureau's plans to modernize its geographic resources. Chapter 4 examines a particularly crucial piece of the Bureau's plan for 2010: implementation of the American Community Survey and, with it, elimination of the census long form, and we suggest critical research priorities for the survey. Chapter 5 explores enumeration and data-processing methods for the 2010 census, focusing primarily on the Census Bureau's plans to use portable computing devices for nonresponse follow-up enumeration. Our discussion on enumeration methods also provides commentary on the need for attention to the Census Bureau's handling of special places and group quarters and to the enumeration of hard-to-count populations. We also describe two elements of the data-processing stage that emerged as concerns in the 2000 census: unduplication and imputation. Chapter 6 describes the Census Bureau's attempts to more effectively model its technical infrastructure—work that is critical to making sure that all the activities and plans described in the preceding chapters are efficient and supported by a reliable technological base. Finally, Chapter 7 briefly discusses the need for coverage measurement and evaluation in the 2010 census.

In Part III, we return to high-level themes of 2010 census planning. Chapter 8 discusses the Census Bureau's research and evaluation program, noting the need for a comprehensive evaluation program and for better exploitation of resources such as the Master Trace Sample. Chapter 9 discusses priorities for census testing, outlining what we believe to be crucial elements of a proof-of-concept test in 2006 and a dress rehearsal in 2008.

CHAPTER 2

The General Plan for the 2010 Census

IN ITS EARLY PLANNING FOR THE 2010 CENSUS, the Census Bureau has, to its credit, suggested a general approach that promises much more than a mere incremental improvement of the census performed 10 years earlier. If carried through as planned and if all of its elements are fully realized, the 2010 census will result in the most dynamic change in census-taking since the 1970 switch to an enumeration conducted principally by mail rather than by ringing doorbells or knocking on doors. The success of the Bureau's vision—an extensively reengineered 2010 census—is, however, highly contingent on the completion of much difficult work in the next few years.

As context for discussion of the developing plans for the 2010 census, we review the basic steps involved in performing a decennial census in Section 2–A. We then briefly review the major milestones and problems of the 2000 census and its planning process (Section 2–B) before outlining the major initiatives of the Census Bureau's general plan for the 2010 census (Section 2–C). Finally, in Section 2–D, we express and elaborate on our belief that, based on the information that has been provided to us, many of the Bureau's plans for 2010 are at serious risk of failure if certain deficiencies are not corrected. This is our greatest concern in this report and it will be discussed in detail in the chapters that follow.

2–A BASIC STEPS IN THE DECENNIAL CENSUS PROCESS

Conducting a decennial census of the United States presents massive logistical challenges on many levels. It has been said that the fielding of the 2000 census—with more than 860,000 short-term employees serving as enumerators—constituted the "largest peacetime civilian mobilization" in American history (U.S. Department of Commerce, Office of Inspector General, 2000:3). To motivate discussion of the proposed changes for the 2010 census process, it is helpful first to consider the basic steps that characterize the modern decennial census. We have listed these basic steps in Box 2.1 and describe them in more detail (with particular reference to their implementation in the 2000 census, as appropriate) in the remainder of this section.

2–A.1 Preparation

The ultimate quality of a census depends critically on successful completion of a number of preparatory steps. First, the Census Bureau must establish an organizational structure (1) as it mobilizes for the count, from the definition of staff roles at Census Bureau headquarters in Suitland, Maryland, to the establishment and operation of hundreds of temporary local census offices (LCOs). The basic high-level organizational structure of the Census Bureau is described in Box 2.2. In addition to the human organizational structure, census planners must also piece together the broader *technical infrastructure* of the census—the amalgam of people, computer hardware and software systems, and telecommunication networks that will be used to support all aspects of the census process.

Of the initial preparations that must be made for a census, the development of an address list (2) is arguably the most crucial. Recent decennial censuses have relied heavily on the delivery and return of questionnaires by mail for most of the population, and the quality of mail census operations relies on the strength

Box 2.1 Basic Steps in Conducting the Decennial Census

Preparation

1. Establish organizational and technical infrastructure
2. Develop an address list
3. Determine questionnaire content
4. Design questionnaires and mailing materials
5. Develop enumeration procedures and assign to areas of the country
6. Design plan for coverage evaluation
7. Establish advertising and outreach programs

Taking the Count

8. Deliver questionnaires
9. Follow up with nonresponding households
10. Conduct coverage evaluation operations

Data Processing

11. Capture data from completed questionnaires
12. Apply editing, imputation, and unduplication procedures
13. Tabulate and release data

Research (before, during, and after census)

14. Test proposed operations and procedures
15. Evaluate operations and procedures
16. Design and conduct experiments and trials

of the address list. The address list, combined with other geographic resources, is also essential to the accurate tabulation of census results. In 2000, the Census Bureau's address list—the Master Address File (MAF)—was constructed by augmenting the preserved 1990 address list with inputs from additional sources. The development of the 2000 MAF is described in detail in Section 3–A.1.

In 2000, as in previous censuses, a short-form questionnaire asking for basic data items was administered to most households, while a census long form asking the basic data items along with many other socioeconomic and demographic questions was administered to an approximate 1-in-6 sample of households. Hence, determination of questionnaire content (3) is an impor-

Box 2.2 Organization of the Census Bureau

The Census Bureau is headed by a director, appointed by the president with the advice and consent of the Senate. The directorship is a political, not a fixed-term, appointment. In addition to a deputy director, the Bureau's current organizational chart also includes two principal associate director positions, one of which serves as chief financial officer, and the other of which is designated as principal associate director for programs (the principal associate director positions are currently not filled).

Under these top levels of management, eight associate directors oversee particular aspects of the Census Bureau's operations. Of these directorates, two are particularly key to decennial census operations:

- *Decennial Census:* The associate director for decennial census oversees all decisions for census planning, budget, and operations, and all funding for decennial-census-related activities is coordinated under this directorate. Under this directorate, the Decennial Management Division is responsible for planning and coordination as well as liaison with oversight groups such as the U.S. General Accounting Office, the U.S. Department of Commerce Inspector General, and congressional committees. The Decennial Statistical Studies Division develops the statistical methodology for use in census operations, while the Decennial Systems and Contracts Management Office is the primary contracting arm for decennial census operations (including external contracts for such operations as questionnaire printing and data capture). Finally, the Geography Division maintains the Master Address File and TIGER geographic database and administers the Bureau's other cartographic activities.
- *Field Operations:* The associate director for field operations oversees the Field Division, which is charged with conducting the actual field enumeration for the census and other survey programs. This directorate also oversees the Bureau's permanent National Processing Center in Jeffersonville, Indiana, and 12 regional offices; during conduct of the decennial census, hundreds of temporary local census offices also come under this directorate's purview.

Other directorates include divisions that play key roles in the decennial census:

- *Demographic Programs:* Until recently, the American Community Survey was administered by the Demographic Surveys Division under the Demographic Programs directorate; this authority was transferred to the Decennial Management Division. A key part of the Demographic Programs directorate is the Population Division, which is responsible for the production of postcensal population estimates between census years.
- *Methodology and Standards:* The Methodology and Standards directorate is home to the Planning, Research, and Evaluation Division, which administered the suite of evaluations of census operations and data quality in 2000 (see Section 8–A). Additional expertise is provided by the Statistical Research Division and the Computer Assisted Survey Research Office, which assists with the development and testing of electronic questionnaires.

Box 2.2 (continued)

Basic administrative support for the Census Bureau and decennial census activities is provided by three directorates: Finance and Administration, Information Technology, and Communications. The eighth directorate in the Census Bureau hierarchy is Economic Programs; while it is not involved in decennial census operations, it administers the Census Bureau's extensive portfolio of business and economic surveys, including the stand-alone Economic Census.

SOURCES: Thompson (2000); Census Bureau organizational chart available at http://www.census.gov/main/www/m-img/orgchart.jpg [2/23/04].

tant part of census preparation. The exact set of questions has varied in the past, depending in part on legal and regulatory requirements on the collection of data, as has the mix of questions between the short and long forms. In the 2000 census, for instance, marital status was shifted to the census long form along with such questions regarding housing as number of rooms and amount spent on rent (see National Research Council, 2004:App. B). Once questionnaire content has been finalized, the actual questionnaires and mailing materials must be developed and tested (4). Even small changes on questionnaires can be consequential, and so the exact phrasing and layout for questions such as race and Hispanic origin are extensively debated. For 2000, the question on Hispanic origin was moved to immediately precede the race question and—for the first time—census respondents could choose to identify themselves as belonging to more than one racial category.

Critical choices then involve determining exactly which enumeration procedures (5) will be applied to which areas of the country. Most of the population—some 82 percent in 2000—lives in areas with predominantly city-style (street number and name) addresses, and so is assigned for enumeration via mail (questionnaires are sent by mail and are expected to be returned by mail). But mailout/mailback was just one of nine types of enumeration areas (TEAs) used in the 2000 census, with a variety of other methods being applied to count the remaining 18 percent of the population. Among these other options are update/leave, used in predominantly rural areas, in which enumerators drop off questionnaires at specific housing units and update address information as possible; it is hoped that residents will then return the form by mail. In 2000, the Census Bureau applied update/leave methods in select portions of some urban areas in which it was thought that there might be problems with mail delivery. Other approaches used in different TEAs include list/enumerate (updating address records and collecting questionnaire responses in a single enumerator visit) and special operations for the military and remote areas in Alaska. In addition to the nine designated TEAs, separate operations are deployed to count the population residing in group quarters (those living in places such as hospitals, dormitories, and prisons) and the population served by facilities like soup kitchens and shelters.

As we will discuss further in Section 2–B, the basic census step that garnered the most attention and focus in preparing for the 2000 census was designing a plan for coverage evaluation (6). Coverage evaluation programs are intended to use independent measures to estimate the accuracy of the census count, both in the aggregate and for specific demographic groups. Coverage evaluation plans in 1990 and 2000 centered around an independent postenumeration survey, though demographic analysis estimates (essentially, updating population counts by adding births and immigrants and subtracting deaths and emigrants) were also used as a corroborative measure. A final piece in the preparation for the census count is establishment of advertising and outreach programs (7) to facilitate cooperation in the census. The 2000 census departed from past programs that relied on donated commercial time; the Bureau was instead more aggressive in marketing and raising public awareness of the census.

2–A.2 Taking the Count

The actual conduct of the census is done in two operational phases: delivery of questionnaires (8), either by mail or in person depending on the type of enumeration area, and visits by enumerators to follow up with nonresponding households (9). Of course, this basic breakdown masks a great deal of complexity, as each of these broad steps involves significant planning and constant management.

Particularly complicated is the process of enumerator follow-up, which itself had two phases in the 2000 census. The first stage, nonresponse follow-up (NRFU), sent enumerators to each household that did not return a questionnaire to either collect questionnaire information or ascertain the status of the address (e.g., whether it was a vacant unit or nonresidential). Enumerators were instructed to collect questionnaire information whenever they could obtain an interview, even if the respondents indicated that they had already mailed back a form. In 2000, NRFU lasted 8 weeks, between April and June, and processed a total workload of 41.7 million addresses (National Research Council, 2004). The second stage of follow-up in 2000, coverage improvement follow-up (CIFU) made further checks on 8.9 million housing units, the majority of which had been labeled vacant or nonexistent in the NRFU operation. For mail returns, a coverage edit follow-up (CEFU) operation was conducted by telephone in order to resolve discrepancies between the number of residents claimed on the form and the number for which characteristics were actually reported (the 2000 questionnaire limited households to providing detail on 6 members). The CEFU phase did not involve the use of field enumerators in the event that telephone contact failed.

After census field operations have concluded, coverage evaluation operations (10) are initiated. For coverage evaluation using a postenumeration survey (as in 2000), the basic goal is to conduct a survey that is completely independent of the census operations and that obtains information about people who lived in particular households on Census Day, April 1. Thus, the postenumeration survey must be conducted as quickly as possible after Census Day (to minimize reporting errors by respondents) but not so close that census and coverage evaluation enumerators are simultaneously working on getting responses from households (which could violate the assumption of independence).

2–A.3 Data Processing

Particularly crucial segments of the census technical infrastructure come into play once questionnaire delivery begins and questionnaires are returned. To begin, the data must be extracted from the paper questionnaire, in a process known as *data capture* (11). For the 2000 census, the Census Bureau took the step of working with outside contractors and incorporating optical character recognition (OCR)—electronically parsing handwritten responses on paper questionnaires scanned into a computer. [The 1990 and earlier censuses had relied on optical mark recognition (OMR), in which the computer is intended to recognize marks only in particular locations on the questionnaire (e.g., a filled-in circle to indicate a particular response category).] Data were input by hand from scanned images of the questionnaire, as necessary, if the OMR and OCR methods failed for some reason.

Once captured from the questionnaires, the census data are then processed. In particular, editing, imputation, and unduplication procedures (12) are applied to the data. Edits range from the very simple (e.g., calculating age if date of birth is given but age is not reported) to very complex, such as assessing the consistency of reporting of various types of employment information on the census long form. Imputation routines can be used to fill in values for missing data items on otherwise complete household data records or to fill in values for entire persons if their information is not included on the census return. For the past several censuses, the Bureau has relied on "hot-deck" methods that fill in responses using a pool of possible responses, drawn roughly from questionnaires in nearby households. Finally, the data are unduplicated to the extent possible, in terms of both persons and housing units. The 2000 census relied on a primary selection algorithm (PSA) to determine a single response if multiple data records (questionnaire results) were reported for the same housing units. Unduplication experience in 2000 is described further in Section 2–B.

Finally, when the data have been processed, they are ready for tabulation and release (13). By law, census counts by state must be reported to the president by 9 months after Census Day for the purpose of reapportioning the U.S. House of Representatives. Under the now-traditional April 1 Census Day schedule, this means generation of apportionment population counts by December 31 of the census year. Census law also mandates that data for legislative redistricting must be issued within 1 year of Census Day.

2–A.4 Research

A final broad set of basic census steps relate to research. These steps differ from the ones previously listed because they deviate from the rough chronological flow of the census process. In principle, census research should occur before, during, and after the census.

In terms of testing proposed operations and procedures (14), it has become the norm for the last major test before the census to be a dress rehearsal 2 years prior. Each census is also accompanied by a suite of evaluations of census operations and procedures (15). For the 2000 census, the Bureau developed a program of 91 formal evaluation studies related to census operations (reduced from an initial total of 149). These operational evaluations were complemented in the 2000 census cycle by three separate waves of research studies connected to the Accuracy and Coverage Evaluation program, each of which informed a decision on the use of adjusted census figures (see Section 2–B).

Finally, each recent census has been accompanied by several formal experiments (16), generally conducted at the same time as the census but focusing on major processes or operations that are not yet ready for full implementation in the census. In 2000, major experiments included a study of respondents' privacy attitudes and willingness to divulge Social Security numbers on the census form, a response mode and incentive experiment that gauged differences in response when a reward (specifically, a prepaid telephone calling card) was offered for submitting census information via the Internet or telephone, and an alternative questionnaire experiment testing some question wording and formatting options. Significantly for 2010 planning, a major experiment of the 2000 census was fielding of the Census 2000 Supplementary Survey, conducted as an experiment to judge whether it was feasible to deploy this survey and run decennial census operations simultaneously. The Census 2000 Supplementary Survey was the first nationwide prototype of the American Community Survey, a centerpiece of the Bureau's emerging plans for 2010.

2–B PLANNING AND CONDUCTING THE 2000 CENSUS

The final report of the Panel to Review the 2000 Census includes a rich and detailed history of the evolution of the 2000 census design and a review of its operations (see especially National Research Council, 2004:Ch. 3, 5). Given the material in that volume, we do not intend to repeat a comprehensive history of the 2000 census in these pages but rather to review major milestones and problems in the 2000 census planning process.

By many accounts, the planning process for the 2000 census was fraught with risk and ultimately chaotic. In 1997, the U.S. General Accounting Office (GAO) added the 2000 decennial census to its short list of high-risk programs, citing failure to finalize and justify to Congress a basic census design. In 1990, "the most expensive census in history produced results that were less accurate than those of the preceding census" (U.S. General Accounting Office, 1997:142), and GAO worried that the cycle could repeat—that billions of dollars invested in the 2000 census might produce results that were not of demonstrably better quality than those of the 1990 census.

To a great extent, 2000 census planning was dominated by ongoing debate over the role of sample-based methods in the census. The Census Bureau's initial plans in the mid-1990s relied heavily on sampling in two respects: first, the Bureau would focus nonresponse follow-up efforts on a sample of households that did not return their census forms and, second, the results of a major postenumeration survey would be used to adjust census totals to reflect estimated undercount. The former application became known as sampling for nonresponse follow-up (SNRFU) while the planned postenumeration survey was known as Integrated Coverage Measurement (ICM). The Census Bureau's vision at the time was a "one-number census," producing a single set of population estimates using ICM results rather than a dual-track approach of publishing both adjusted and unadjusted figures.

Both applications of sampling—and, with them, the overall plan for the 2000 census—would ultimately be shaped by intervention first by Congress in 1997 and then by the U.S. Supreme Court in 1999. The 1997 legislation (Public Law 105-119, §209) required the Census Bureau "to plan, test, and become prepared to implement a 2000 decennial census, without using statistical methods, which shall result in the percentage of the total population actually enumerated being as close to 100 percent as possible." While it did not expressly prohibit SNRFU or ICM, the legislation did signal Congress's insistence that the Census Bureau be prepared to enact a more traditional census model. In addition, the legislation dispensed with the "one-number census" vision by mandating that apportionment counts, redistricting files, and other 2000 census products contain "the number of persons enumerated without using statistical methods" as well as those added or subtracted using sample-based techniques.

The 1997 legislation also required expedited judicial review of cases filed by parties who believed themselves aggrieved by the use of methodologies based on sampling. Two major challenges arose as a result and were ultimately consolidated for hearing by the U.S. Supreme Court, setting the stage for the final decision on the design of the 2000 census. In January 1999, the Court issued a 5–4 ruling that the Census Act (Title 13 of the U.S. Code) prohibited the use of sample-based methods in tabulating population counts for reapportioning the U.S. House of Representatives (*Department of Commerce v. United States House of Representatives*, 525 U.S. 316). The Court's ruling stopped short of declaring sampling unconstitutional in the census process, and it left open the possibility of adjustment of census data for other purposes (including legislative redistricting) based on a postenumeration survey. But, despite these remaining ambiguities, the Court's ruling forced the Census Bureau to abandon ICM and revamp the 2000 census plan little more than a year from April 1, 2000, the census target date. ICM was ultimately replaced by a smaller postenumeration survey, the Accuracy and Coverage Evaluation (ACE) Program.

The delay in finalizing a census design meant that the 1998 census dress rehearsal was a major experiment rather than a true dress rehearsal. Instead of permitting a full test of finalized plans and systems, the dress rehearsal conducted in spring 1998 was primarily a field comparison of competing basic designs. The original sampling-based (ICM and SNRFU) framework was tested in Sacramento, California; a traditional census with a postenumeration coverage survey was tested in Columbia, South Carolina; and a hybrid approach was fielded in Menominee County, Wisconsin.

In addition to the delay in finalizing a general design, unanticipated difficulties affected other parts of the census process. The Census Bureau had assumed that updating its Master Address File with information from the U.S. Postal Service and local and tribal governments would produce an address list of adequate currency. But—late in the census process—the Bureau concluded that gaps remained and, between January and May 1999, census field staff performed an extensive (and expensive) canvass of the address list. During the actual conduct of the census, further evaluation of the address list suggested that the list had sizable levels of duplicate housing unit addresses, leading to an ad hoc operation to screen potential duplicates for further examination and possible removal or reinstatement into the census (Nash, 2000). As discussed in National Research Council (2004)—and as we discuss in Chapter 3—the Census Bureau's strategy for using multiple sources to create the Master Address File for the census was well intentioned but not well executed.

The computer systems used to support census operations were developed swiftly and often lacked adequate testing. The Census Bureau also, for the first time, relied heavily on outside contractors for assistance in parts of the census process, notably data capture (scanning paper questionnaires to recover data in electronic format). Although the outside contracting ultimately proved successful, it did create some concern; for instance, the color chosen for the background of the census questionnaires was later found to be near the borderline of acceptability as a background for automated optical scanning, and slight variations in tint incurred during printing could have caused the scanning system to reject questionnaires (Titan Corporation, 2003:19). Information systems also led to a brief public embarrassment when every letter mailed out in the nationwide advance notification—letting the public know about the imminent arrival of census questionnaires—was misaddressed, with an extra digit prepended to every street number. Fortunately, the postal bar code on the letter contained correct information and the U.S. Postal Service was able to deliver the letters (Prewitt, 2000; Norman, 2000). In addition, the Bureau had planned to implement a second questionnaire mailing to nonresponding households—a survey practice well known to improve response rates, and one that previous National Research Council panels have strongly endorsed. However, as the census neared, the Bureau determined that its contractor would be unable to turn around the printing and mailing of follow-up questionnaires on a sufficiently quick timetable, and plans for the second questionnaire were dropped—to the likely detriment of final mail return rates.

As the 2000 census progressed, some hints of impending problems and concerns about the Census Bureau's enumeration methodologies began to emerge, although the full extent of any problems would not be known until well after 2000. Press coverage of the census highlighted increased concern over privacy and the perceived intrusiveness of some census questions, particularly those on the census long form.[1] Though countered by comments from other politicians and Census Bureau officials who stressed the importance of census long-form information, the publicity spotlighted increasing public concerns about privacy and raised the possibility of low and poor response to the long form. As another portent of things to come, various groups charged with overseeing the census process expressed concern about what would be described by others as a success of the 2000 census: nonresponse follow-up operations were concluded ahead of schedule. Oversight groups worried that local census offices might be closing out their workload too early, not making the fullest effort to follow up with all nonrespondents. The implications of the Census Bureau's strategy to conclude nonresponse follow-up as efficiently as possible and use imputation methods to fill remaining response gaps ultimately led to a U.S. Supreme Court upholding the use of imputation (*Utah v. Evans*, 536 U.S. 452, 2002).

---

[1]Perhaps the highest-profile comments along these lines came from then-presidential candidate George W. Bush and then-Senate majority leader Trent Lott of Mississippi. In particular, Bush was widely quoted as commenting, "I can understand why people don't want to give all that information to the government. If I have the long form, I'm not so sure I want to either" (Gilbert, 2000). See also: "Despite complaints, Lott encourages cooperation on census," http://www.cnn.com/2000/ALLPOLITICS/stories/03/30/census.cnn/ [12/01/03].

State-level population counts for congressional reapportionment were transmitted to the president on schedule in December 2000, but the Supreme Court's 1999 ruling had left open the possibility of statistical adjustment of population counts for other purposes. As the April 2001 deadline for block-level redistricting data neared, a political tug-of-war ensued as to whether the director of the Census Bureau or the secretary of commerce should have the authority to determine the appropriateness of adjustment.[2] A group of senior-level Census Bureau staff—the Executive Steering Committee on ACE Policy (ESCAP)—met regularly to assess the merits of an adjustment and, ultimately, advise the census director.

Subsequently, the Census Bureau concluded on three separate occasions that it could not reliably use the ACE to adjust 2000 census totals. In March 2001, ESCAP determined the presence of a net undercount—both overall and differential among racial groups, at the national level—but recommended against adjustment for purposes of redistricting (a recommendation promptly accepted by the acting census director and approved by the secretary of commerce). The Census Bureau ESCAP recommendation cited difficulties reconciling ACE results with population estimates produced through demographic analysis, as well as concern over possible errors in the ACE process (Executive Steering Committee for A.C.E. Policy, 2001b; National Research Council, 2001a). ESCAP continued deliberation and analyzed further research in advance of an October 2001 decision on adjustment for such purposes as allocation of federal funds. At that point, ESCAP again opted against adjustment (Executive Steering Committee for A.C.E. Policy, 2001a), arguing that the original ACE had overstated the national net undercount by at least 3 million people because it failed to identify many erroneous enumerations (including duplicates) in the census. More Census Bureau research on the ACE continued

[2]Clinton commerce secretary Norman Mineta delegated authority to make the adjustment decision to the census director in early 2000 (65 *Federal Register* 195, 59713–59716)—a move rescinded by Bush commerce secretary Donald Evans in 2001 (66 *Federal Register* 37, 11231–11233). Both versions of the regulation stipulated the formation of the Executive Steering Committee on ACE Policy within the Census Bureau.

through 2002, culminating in a third Census Bureau decision against adjustment, this time in March 2003 and related to adjustment of census data for use in deriving postcensal population estimates. The Census Bureau's final estimates suggested a national net *overcount* of 0.48 percent (1.3 million persons), with continued differential undercount among some racial groups.

Flaws arose in the conduct of the 2000 census, as they inevitably arise in every census; however, the 2000 census was ultimately successful in meeting its statutory deadlines for providing data for reapportionment and redistricting. It was also successful in curbing the trend of past censuses toward lower overall mail return rates, among other accomplishments. That said, the process by which the 2000 census plan developed leaves considerable room for improvement. The final design for the 2000 census was put into place an inadvisably short time before the census had to go into the field. Looking ahead to 2010, both the Bureau and outside observers hope to avoid the risks and bruising consequences of late-formed plans, while at the same time keeping in check the escalating costs of conducting a census of the complex U.S. population.

2–C THE "THREE-LEGGED STOOL" APPROACH TO THE 2010 CENSUS

In the early planning stages, the Census Bureau identified four basic goals for the 2010 census (Waite, 2002; Angueira, 2003b):

1. increase the relevance and timeliness of census long-form data;
2. reduce operational risk;
3. improve the coverage accuracy of the census; and
4. contain costs.

Based on these goals, the Census Bureau developed a general strategy for the 2010 census even as 2000 census returns were still being processed. As first described to the panel at its December 2000 meeting, the Bureau's general strategy for 2010 was likened to a "three-legged stool," predicated on three major initiatives:

- *Modernization of the Census Bureau's geographic resources.* The Bureau's Master Address File (MAF) and its geographic database (Topologically Integrated Geographic Encoding and Referencing System, or TIGER) will be updated so that they will be consistent with coordinates derived using global positioning systems (GPS). The intent is to save field time and costs as well as improve data accuracy.
- *Implementation of the American Community Survey (ACS).* This proposed sample survey will collect data on the same social, economic, and demographic variables included in the current census long form, but will do so on a rolling continuous-time basis. As in earlier censuses, the census long form was administered in 2000 to a sample of households (1 in 6) while most households received the short form. Full ACS implementation will permit the 2010 census to be conducted using only the short form. It is hoped that this change will facilitate the collection of information through the Internet and simplify data capture from census forms returned by mail.
- *Early integrated planning.* To the extent possible, census plans will be finalized early to enable effective testing in the years leading up to the census. It is hoped that this early planning will make the pre-2010 census tests more useful and informative as well as forestall a costly end-of-the-decade crunch in finalizing census operational plans.

An immediate adjunct to this three-pronged strategy is the incorporation of new technology in the census process. In particular, the Census Bureau's emerging 2010 census plans take advantage of a short-form-only census by including the following additional components:

- *Multiple response modes.* Simplifying to a short form would make completion of the census form easier and quicker—and more tractable for administering to respondents electronically. Hence, it is anticipated that the mailout/mailback component of the census would be heavily augmented with enumeration through use of the Internet and possibly interactive voice response via the telephone.

- *Portable computing devices (PCDs).*[3] Nonresponse follow-up field work will make use of portable computing devices for communication of assignments, computer-assisted interviewing, and data capture. With an enhanced MAF/TIGER database, the Census Bureau also anticipates that PCDs equipped with GPS receivers will allow interviewers conducting nonresponse follow-up to pinpoint the location of their assigned housing units and, possibly, to optimize their navigation from one assignment to another.

Final specifications and detailed plans for the above design remain to be developed.

2–C.1 Relation of the 2010 Plan to Basic Census Processes

In the context of the basic census steps described in Section 2–A, the major focus of activity in the planning of the 2000 census was in step (6), designing a plan for coverage evaluation, and the effects of that general design on other parts of the census process. By comparison, the Census Bureau's emerging plan for 2010 effects its major change in step (3), questionnaire content, by eliminating the census long form, replacing it with the ACS, and implementing a short-form-only census. This change, in turn, causes a ripple effect throughout other parts of the process—for

[3]The Census Bureau has used the term "mobile computing devices"—or, more frequently, the acronym MCD—to describe the small computers planned for 2010 field data collection. However, the choice of MCD as a label is unfortunate because the acronym conflicts with that for "minor civil division," a long-standing concept of census geography referring to the subcounty (township) divisions that are functioning governmental units in several Midwestern and Northeastern states. Since issuance of the panel's second interim report, some Census Bureau materials have referred to the devices as PDAs—"personal digital assistants"—or handheld devices, in usage consistent with current Census Bureau expectation that the devices will be similar to current Palm Pilot-class devices. We have adopted the compromise term "portable computing device" (PCD).

instance, making the Bureau's overall organizational and technical infrastructure more complex through the introduction of a parallel data collection effort in the ACS.

The emerging 2010 plan also portends major change in usual approaches to step (11), data capture from completed questionnaires. The short-form-only decennial census should ease the difficulty and speed the completion of data capture (although, of course, systems will still need to be in place to capture data from the long-form-like ACS). Though the bulk of census information is still likely to be obtained through mailed-out questionnaires being returned by mail, increased use of the Internet and possibly telephone will allow for automated data capture. So, too, will information collected in nonresponse follow-up by enumerators using PCDs—with the added challenge that data from PCDs may be transmitted directly from thousands of enumerators to census headquarters without filtering through data capture centers.

As we will discuss in other parts of this report, the emerging 2010 census plan also includes steps that could affect other basic steps in the census process. These include, for instance, attempts at more rigorous modeling of the underlying census technical infrastructure (Chapter 6) and the possibility of real-time unduplication efforts during the actual conduct of the census (Chapter 5). Another major component of the Bureau's plans—the MAF/TIGER Enhancements Program—suggests changes in the basic development of the census address list. However, as we will argue later in this chapter and in Chapter 3, much greater focus on address development is needed as the 2010 census plan proceeds.

2–C.2 Relation of the 2010 Plan to the 2000 Census

The kernel of the 2010 census—the set of three major initiatives—began to take shape in the Census Bureau's plans in the late 1990s and in 2000, and was presented to the panel in December 2000. Accordingly, the first thing that must be acknowledged regarding the Census Bureau's emerging plan for 2010 is that it did not and indeed could not arise directly from empirical evidence drawn from evaluations of the 2000 census. The earliest 2010 planning efforts could draw from operational and anecdotal evidence as the 2000 census was fielded, but formal evaluation work on the 2000 census was long delayed (with the large body of this work not publicly available until fall 2003), and the complicated picture of coverage measurement and related research only took shape in the various waves of ACE research from 2001 to 2003.

The list of major problems and concerns arising from the 2000 census includes flaws in the Master Address File, duplication of both persons and housing units, a widening gap between short-form and long-form response rates, and problematic handling of special places and group quarters. The tripartite core of the 2010 census plan reflects some of these concerns (particularly the issue of nonresponse to the long form) but was not directly tailored to deal with them. By noting this fact, we do not imply that the 2010 plan is unrelated to census experience; the plan reflects the four basic goals articulated by the Bureau, which in turn reflect long-standing problems in the census process. Nor do we suggest that all 2010 census planning should have been postponed until all evaluations were completed; as we argue in this report, it is to the Bureau's great credit that planning began as early as it did. Instead, what we suggest is that—because the 2010 plan did not evolve directly from research and evaluation results—it is vitally important for the Census Bureau to clearly articulate the connection between planned 2010 programs and the evidence available from the 2000 census.

2–C.3 Planning Milestones of the 2010 Census

The Census Bureau anticipates four major census tests prior to 2010 in order to try out new procedures and finalize program plans. The chronology of these tests and other milestones in the planning process for 2010 are shown in Table 2-1. In 2003, a national sample was asked to participate in a test of possible response modes (e.g., mail, Internet, and telephone) and of rewordings of the questions on race and Hispanic origin. The 2003 test was administered only by mail and did not involve an active field deployment of enumerators to conduct follow-up questioning. The 2004 Census Field Test will cover a wider range of census operations, including field follow-up, in predetermined sites in Georgia and New York.[4] The 2004 field test should be the first major test of the Bureau's plans for using portable computing devices. In January 2003, the Census Bureau announced an Overseas Enumeration Test to be fielded in 2004, a test intended to gauge the response of U.S. citizens living in France, Kuwait, and Mexico to outreach and marketing efforts (U.S. Census Bureau, Public Information Office, 2003). The Census Bureau has designated an as-yet unspecified test in 2006 as its "systems test," focusing on general and reengineered technical systems. Finally, a full-fledged dress rehearsal is to be conducted in 2008. The Census Bureau hopes that avoiding a late-decade crush in designing census plans will make the 2008 exercise a true rehearsal rather than a late experimental test, as was the case with the 2000 census dress rehearsal.

2–C.4 Status of the 2010 Census Plan

To date, the Census Bureau has prepared and shared with the panel three draft documents related to the general 2010 census plan: a baseline design consisting of a bulleted list of intended design features (Angueira, 2003b); a "plan for the plan" that outlines goals and objectives for the 2010 census (Angueira, 2003a); and a risk management plan (Decennial Management Division, 2003). We understand the draft nature of the documents and so will directly quote from them as necessary but not to a great degree. In addition, the panel has received a report sketching costs of the 2010 census plan that has been disseminated more widely to census oversight and advisory groups (U.S. Census Bureau, 2001a).

The early timelines established for 2010 census planning have already been affected by the federal budget process. Extensive delays by Congress in passing a budget for fiscal 2003 (relying for several months on the passage of continuing resolutions, to keep federal spending at current levels) forced a delay in full implementation of the ACS. Originally intended to start full op-

---

[4]Plans for a 2004 test in Lake County, Illinois, were dropped following release of the Bush administration's budget requests for fiscal 2004 (Lowenthal, 2003b).

Table 2-1 Planned Testing and Development Cycle for the 2010 Decennial Census, Assuming a Short-Form-Only Census

Year	Census Activity
2002	Begin planning and develop methods for 2004 Census Field Test
2003	Conduct 2003 National Census Test, a survey administered by mail but offering multiple response modes (mail, telephone, Internet) and rewording of race and Hispanic origin questions
2004	Conduct Census Field Test, emphasizing use of portable computing devices, in selected sites in New York, Illinois, and Georgia; conduct Overseas Enumeration Test in France, Kuwait, and Mexico
2005	Analyze results and refine methodology
2006	Conduct National Census Test, involving prototype technical systems; (possibly) conduct second overseas census test
2007	Analyze results and refine and integrate systems and methods
2008	Dress rehearsal
2009	Begin to implement operations
2010	Conduct census

SOURCES: Waite (2002); U.S. Census Bureau, Public Information Office (2003).

erations in 2003, the ACS is now scheduled to start in full in the last quarter of fiscal 2004.

As Congress worked on fiscal 2004 appropriations bills, the two houses proposed markedly different budget totals for the Bureau. The House version conformed to the administration's request and included specific appropriations for the ACS, but the Senate mark imposed a major cut in funds for periodic censuses and related programs (the House approved \$436,053,000; the Senate cut the total to \$369,067,000). In response, the Census Bureau scaled back plans for its 2004 test, dropping the planned Lake County, Illinois, test site and trimming some components from original plans, and public comments by the Bureau warned that passage of the Senate version might lead to further cuts in testing and a slowdown of the MAF/TIGER modernization program. In December 2003, House and Senate conferees proposed an omnibus bill, consolidating several separate appropriations bills, that acceded to the House funding levels for the 2010 census, save for a cut from $8.6 million to $3.6 million in the allotment for "operational design strategy" (H. Rept. 108-401). The bill allots $107,090,000 for 2010 census planning, $83,310,000 for the MAF/TIGER Enhancements Program, and $64,800,000 for the American Community Survey in fiscal 2004. The bill was approved by Congress in mid-January 2004.

The Census Bureau requested a $180 million increase in fiscal 2005 funds for 2010 census activities.[5] This increase from fiscal 2004 totals was included in the Bush administration's budget message for fiscal 2005, which notes that the budget increase includes a first full year of funding for the American Community Survey (*Budget of the United States Government, Fiscal Year 2005*, p. 77).

2–D REENGINEERING THE 2010 CENSUS: A PROCESS AT RISK

The Census Bureau has advanced an ambitious vision for the 2010 decennial census, and—as our previous reports and the balance of this report suggest—the panel strongly supports the major aims of the plan. The implementation of the ACS, for example, and with it the elimination of the long form from the decennial census process is a very good idea; the Bureau's geographic databases are in dire need of comprehensive update; and the implementation of new technologies in census-taking is crucial to maintaining an accurate count. There is thus much to like about the emerging plans for the 2010 census, and we strongly support these efforts toward a modernized and improved census in 2010. To this end, the Census Bureau's focus on planning early in the decennial cycle is highly commendable.

Based on the information made available to us, however, the panel finds that some of the planned-for innovations in the reengineering of the 2010 census are at considerable risk of failure or partial failure. Many of the problem areas that we see

[5]See "EXECUTIVE SUMMARY—FY '05 Budget" issued by the Census Bureau, available at http://www.census.gov/Press-Release/www/releases/archives/miscellaneous/001675.html [2/23/04], for additional detail on the Census Bureau request.

in the emerging 2010 census plan stem from what we believe to be a serious disconnect between research and operations in the census process. Put another way, the Census Bureau's planning and research entities operate too often at either a very high level of focus (e.g., articulation of the "three-legged stool" concept for the 2010 census) or at a microlevel that tends toward detailed accounting without much analysis (e.g., the program of planned evaluation reports of the 2000 census; U.S. Census Bureau (2002a)). What is lacking is research, evaluation, and planning that bridges these two levels, synthesizing the detailed results in order to determine their implications for planning while structuring high-level operations in order to facilitate meaningful detailed analysis. Justifying and sustaining the 2010 census plan requires both research that is forward-looking and strongly tied to planning objectives, and rigorous evaluation that plays a central role in operations rather than being relegated to a peripheral, post hoc role.

The Census Bureau still needs to do a great deal of work to develop a strong research and evidentiary base for the general 2010 census plan, carefully assessing operational data from the 2000 census to guide planned practice for 2010 and fully exploring the potential of new tools for evaluation (such as the Master Trace Sample containing results of all census operations for a limited national subset). Looking ahead, much work also remains in integrating and mapping the logical and technical infrastructures of the entire census process, and in developing a rigorous and timely testing program for new census systems and techniques. The consequences of failing to develop a strong research base for the 2010 census are serious. They entail:

- repeating past census processes that may be inefficient or suboptimal,
- conducting a census with methods that are out of step with the dynamics of the population it is intended to count,
- making limited technological innovations that may not match real needs, and
- marking a flawed beginning for 2020.

The Census Bureau, acutely aware of the risks and problems that resulted from extended delay in finalizing plans for the 2000 census, has identified the reduction of risk as one of the key goals for the 2010 census plan and has drafted a risk management plan that identifies 14 perceived risks related to the 2010 census (Decennial Management Division, 2003); these are listed in Table 2-2. The draft risk management plan is a useful start but it needs extensive and continuous revision. In approach, it is perhaps overly formulaic in judging the severity of risks and may make it too easy to consign moderate-risk items to an amorphous "tracking list," rather than performing the research necessary to more fully evaluate the risk and determining potential remedies. Of the 14 risks identified by the Bureau, 9 are designated as actionable by placement on the tracking list. In our assessment, this severely underestimates the risks related to the technical infrastructure of the census (e.g., estimating the risks of "systems not performing or obsolete" and "stovepipe systems that are not interoperable"[6] to be low). As we discuss in greater detail in Chapter 6, although the Bureau is working on steps to effectively model its infrastructure, its success in using that modeling capability to the full extent (in order to abate risk) is not a foregone conclusion. Furthermore, the Census Bureau's risk management plan does not identify risks associated with the major components of the general census design, such as the ACS and the MAF/TIGER modernization, and the plan's soft-pedaling of the risk associated with the lack of integration of these components is puzzling.

As the Bureau's risk management plan notes, 2010 census reengineering faces two paramount risks: first, that the final design of the census (particularly the role of the ACS or a census long form) will be determined late in the process and, second, that funding will be inadequate or at unpredictable levels. These risks differ in nature from most of those discussed above in that they are largely external to the Census Bureau, since funding decisions are made by the administration and, ultimately,

[6]Stovepipe systems are named such because they are vertically—but not horizontally—integrated. That is, they can be elaborate systems that perform one set of functions, but they are unable to interact and share information with other systems in an enterprise's broader architecture.

Table 2-2 Census Bureau Listing of Perceived Risks in 2010 Census Planning

Risk Description	Perceived Level[a]	Strategy[b]
Late changes to final census design	High	Tracking List
Dress rehearsal experimental, not true rehearsal	Medium	Tracking List
Alienate key stakeholders, including advisory groups and state, local, and tribal governments	High	Containment
Change in or lack of funding	High	Contingency
Low mail response	Medium	Contingency
Outsourcing may not meet census needs	Medium	Tracking List
Systems not performing or obsolete	Low	Tracking List
Field infrastructure (PCDs) not ready	Low	Tracking List
Lack of integration of design components (ACS, MAF/TIGER Enhancement, Early Planning)	Low	Tracking List
Failure to maintain systems security	High	Containment
Concern over privacy and confidentiality jeopardizes data collection	Low	Tracking List
Lack of human, financial, schedule, or material resources	High	Contingency
Staff attrition and corresponding loss of census knowledge	Medium	Tracking List
Stovepipe systems that are not interoperable	Low	Tracking List

[a] The Census Bureau derives this categorization by multiplying a score for likelihood of risk (1, 2, or 3, 3 being most likely) by a score for impact of risk if realized (1, 2, or 3, 3 being largest effect). Final scores of 1 or 2 are labeled "low" risk, 3 or 4 "medium" risk, 6 "high" risk, and 9 "very high" risk.

[b] The Census Bureau identifies three possible risk management actions. "Tracking List" means that the risk is kept on a watch list, which is reviewed periodically "to assist the implementation team in 'keeping their eyes on the horizon.'" A "Contingency Plan" "involve[s] preparation for actions to be taken in the event that a nonactionable risk becomes an issue." Finally, a "Containment Plan" "involve[s] specific actions that ***will*** be taken to [r]educe the probability of the risk turning into an issue [and] reduce the negative impact to the project if the risk becomes an issue" (Decennial Management Division, 2003:20).

SOURCE: Decennial Management Division (2003).

Congress. Although the Census Bureau may not have control over funding decisions, it must take a more active role in informing the funding process in at least two ways. First, as we emphasize throughout this report, the Bureau must develop a sound research and evidentiary base for its 2010 census plan, thus making a stronger and more compelling case for sustained long-term funding. Second, the Bureau should be explicit in identifying, articulating, and quantifying the consequences associated with these broad risks—for instance, the impact of reduced funding on the quality of ACS estimates for small geographic areas and population groups. Failure to reach consensus on the role of the ACS in the census process raises the undesirable prospect of reversion to the long form, possibly late in the census process and therefore implemented in a rushed manner, which is likely to result in the same nonresponse and data quality problems as were experienced with the 2000 long form. More significantly, failure to reach closure on census design leaves open the possibility that the detailed socioeconomic and demographic characteristics measured by the current census long form may not be estimated at all in 2010, which would be an unacceptable outcome.

2–D.1 Specific Risk Areas

Beyond these broad, systemic risks, the 2010 census planning process faces many risks of a more specific nature, some of which are acknowledged in the Bureau's draft of a risk management plan and many of which are not. Based on the information known to us, we find that the 2010 census reengineering process may be seriously jeopardized in the following areas, among others (this list of risk areas is ordered for rhetorical flow, not by any estimated level of risk or urgency):

- *Master Address File updating process:* The panel believes this effort to be at severe risk. The Census Bureau's current approach treats MAF updates as "routine maintenance," relying principally on updates from U.S. Postal Service files. Detailed plans for review by local and tribal governments or for address input from ACS enumerators are either nonexistent or have not been shared with the panel. Absent a strong focus on enhancing the MAF in its own right, independent of presumed benefits from linkage to a realigned TIGER database, the 2010 census may be conducted with an address source that has unacceptable levels of housing unit duplication in some areas and coverage gaps in others.

- *Development of census logical and technical architecture:* The Census Bureau has begun efforts toward modeling the logical and technical infrastructure of the decennial census, a development the panel strongly endorses. However, early indications suggest that the modeling of the 2010 census process has focused on detail without any real reengineering taking place. Draft architecture documents are short on analysis and do not decompose census processes to the appropriate level of resolution needed to actually reengineer and to specify hardware/software systems. Should the full potential of architecture modeling not be realized, the consequences are dire: systems may be ill-suited to handle 2010 census process needs, and may fail during actual census operations due to lack of proper testing. Moreover, census hardware and software systems may not fit properly with each other at points of interface—guaranteeing the stovepipe systems that the Bureau's draft risk management plan correctly suggests should be avoided.
- *Realignment of TIGER features:* The initial realignment of TIGER geographic features to be consistent with GPS coordinates may not be completed in time, or change detection for new features after the initial realignment may not be properly performed. These outcomes would have a negative impact on plans for PCD use by field enumerators and would lead to continued errors in geocoding addresses in the census and nonresponse follow-up operations.
- *TIGER database conversion:* The conversion of the MAF/TIGER database from its current homegrown format to a modern, object-oriented computing environment may be slower or more difficult than anticipated. This is particularly dangerous if the Census Bureau attempts the conversion en masse, rather than via a more carefully designed software reengineering process with ample testing.

- *Targeted methods in address list development or enumeration:* Examples of these methods include targeting block canvass to verify address list entries to particular (e.g., high-growth) areas and expansion of update/leave enumeration (where a census enumerator drops questionnaires at housing units, which are then expected to mail them back) in urban areas where mail delivery may not be the most effective means. Failure to implement these methods may result in increased costs and continued problems of enumeration in high-density areas with structures containing multiple (and not well-listed or easily differentiated) housing units.
- *Use of local knowledge:* The Census Bureau's draft risk management plan includes the risk of alienating key stakeholders, including local and tribal governments. Alienation of local authorities is a risk, to be certain, but a more fundamental risk is failure to fully involve them in census design and operations. In addition to serving a critical role as contributors and reviewers of address list information, local and tribal governments can offer guidance in enumerating group quarters and other special populations, tailoring enumeration techniques (e.g., update/leave rather than mailout/mailback) to specific areas within localities, and fostering acceptance and use of the ACS.
- *Enumeration strategies for special, challenging populations:* Special populations requiring specific, targeted approaches and methodologies include immigrant communities, irregular multiunit housing structures, gated communities, colonias along the U.S.-Mexico border, and the homeless. Efforts in this area may be compromised by failure to tailor enumeration techniques to these groups and to clarify the definition and presentation to respondents of the residence rules for the decennial census. Consequences associated with such a failure include poor quality data, failure to meet consumer needs, and continued differential undercount.
- *Special place/group quarters enumeration process:* In 2000, as in previous censuses, procedures for enumerating group quarters were conducted separate from the rest of the cen-

sus process. Group quarters listings were not reconciled with the MAF for households, and little effort was given to the challenges of enumerating different types of group quarters. The enumeration processes for group quarters were not well controlled. Continuing with this approach incurs the risk of duplication, a repeat of the experience in 2000 when whole group quarters were geographically misplaced or miscounted,[7] and ineffective coverage of this small but important population group.

- *Census duplication (both housing unit and person):* Concerns about duplication during the processing of 2000 census records prompted the Census Bureau to mount an ad hoc unduplication effort. Failure to research and improve techniques to identify and correct duplication of both housing units and person records could lead to a repeat of costly, spur-of-the-moment operations in 2010.
- *Imputation:* Though the U.S. Supreme Court upheld the use of imputation in its ruling on Utah's challenge to the 2000 census counts, the debate suggests the need to revisit the techniques used to fill in missing questionnaire items and, in some cases, to impute household size when no information is available for a presumed-occupied unit. The costs and benefits of alternatives to current "hot-deck" methods should be evaluated.
- *Research and development agenda for the American Community Survey:* Though not an immediate threat to the integrity of the 2010 census, failure to implement a strong research and evaluation program for the ACS poses a longer-term risk to the quality and usefulness of the survey and will hinder the ability to fully exploit potential ties between

[7]Many challenges to census counts filed by localities under the Census Bureau's Count Question Resolution program involved geographic misplacement of such facilities as college dormitories and prisons. Most recently, the Census Bureau acknowledged that a 2,673-resident dormitory at the University of North Carolina at Chapel Hill had been double-counted—a highly contentious finding since North Carolina narrowly edged out Utah (by 857 residents) for the 435th and final seat in the U.S. House of Representatives. Utah had failed in two major legal challenges to the 2000 census totals in order to secure a fourth seat in the House (Baird, 2003).

the ACS and programs for producing postcensal population and demographic analysis estimates.

- *Coverage measurement:* As we discuss in Chapter 7, the coverage measurement program in 2010 need not take the same shape as that of 2000, but it is essential that the Census Bureau have the means to determine the accuracy of its count, for the nation as a whole as well as for population subgroups. After the statistical adjustment battles preceding and following the 2000 census, the Census Bureau may be understandably reluctant to take up active debate on coverage techniques for 2010. However, this reluctance incurs the risk that a comprehensive plan for the measurement and assessment of census coverage in 2010 will be deferred until late in the census process.
- *Portable computing devices:* The Census Bureau's plans for the use of portable computing devices (PCDs) in the 2010 census are a particularly exciting part of a reengineered census, but the plans also entail risks associated with the implementation of new technologies. Perhaps most significant is the risk that the Census Bureau may fail to fully understand the direction in which the technology is moving and thus may spend its resources testing devices that are inferior to those that will be available in 2010, in terms of both size and computing capacity. A consequence of this error is that wrong and misleading conclusions may be drawn about the real potential for portable computing devices to improve census data collection. A second risk inherent with the PCD technology lies in making the decision to purchase too early and without fully specified requirements, resulting in the possible selection of obsolete or inadequate devices. Third, and related, is the risk that the Census Bureau may not use the sheer size of its order (perhaps on the order of 500,000 devices) to obtain devices tailored to census needs, as opposed to buying only what is commercially available off the shelf. Finally, given the fact that the principal users of the devices will be the large corps of temporary enumerators (with limited training), there is the risk that Census Bureau PCD development will not take

human factors into sufficient consideration. The full potential of PCDs will not be realized if they are either too simple or too complex.

- *New response technologies:* For example, a repeat of the 2000 experience concerning a second questionnaire mailing for nonresponding households would be highly undesirable.

2–D.2 Mitigating the Risks

Although in this report we criticize certain aspects of the Census Bureau's planning process, we do not want to appear unduly alarmist about it. We are not stating that the 2010 census is irrevocably headed toward a bleak outcome. Our comments are intended, however, to strongly emphasize the need for extensive research and evaluation so that the Census Bureau will be able to stay on course toward achieving its high-level goals for 2010.

Throughout this report, we offer recommendations to improve 2010 census planning and mitigate risks. In brief, some of the steps that we believe will be crucial to the success of 2010 census planning include the following:

- *Recast the 2006 census test as a proof of concept, not a technical test.* The Census Bureau hopes to make its planned 2008 activity as close to a true dress rehearsal as possible, which is certainly desirable. However, that goal implies that the success of the 2006 test is of critical importance: it must provide the basis for definitively answering any remaining experimental questions in order to make the 2008 dress rehearsal a genuine preoperational rehearsal instead of just an extension of earlier tests.
- *Focus on the Master Address File.* Realignment of the TIGER database is undeniably important and long overdue, but it is the quality of the address list that is crucial to the quality of a census. As we argue in Chapter 3, the MAF is in need of much greater attention than it currently receives under the MAF/TIGER Enhancements Program.
- *Use research to bolster the case for the American Community Survey.* Making a stronger case for the utility of ACS-based

estimates (including moving averages) in a variety of applications and further examining the relative quality of ACS and long-form estimates is essential to winning long-term support for the survey.

- *Use logical and technical infrastructure modeling to its fullest extent.* Reengineering to understand, simplify, and minimize inefficiencies and redundancies (both informational and functional) in census systems is essential to break up stovepipe systems and promote integration in census planning.
- *Adapt testing of portable computing devices to anticipate future requirements.* In particular, testing should be done on current high-end devices of the form that are certain to be available, and less expensive, by the time of procurement. In approaching PCDs, the Census Bureau should seek to exploit technology to guide new and better enumeration processes, rather than simply replicating old processes on new tools.
- *Make research and evaluation a centerpiece of census operations, not a peripheral component.* In addition to making better use of extant data resources, the Census Bureau should design its hardware and software systems to facilitate research and allow quick evaluation. In its architecture redesign efforts, the Census Bureau's objective should be an integrated system that provides information for evaluation routinely and in real time—that is, a Master Trace *System*, rather than only a *Sample*.

Part II

Issues of Census Design

CHAPTER 3

Modernizing Geographic Resources

A DECENNIAL CENSUS IS FUNDAMENTALLY an exercise in geography. The root constitutional mandate of the census explicitly links it to the nation's electoral geography, as the census serves as the basis for shifting states' representation in the U.S. House of Representatives every 10 years to match population shifts over the decade. Each new decennial census also offers new perspectives on the nation's civic geography, providing rich information on how and where the American public lives and how the characteristics of small geographic areas and population groups have changed with time. In order to produce this information, the Census Bureau requires a great deal of accurate, raw geographic data—a full and complete address list and a mechanism by which those addresses can be associated with specific locations. Without this raw information, it would be impossible for the census to achieve its goal of counting each resident once and only once and within a precise geographic boundary. As the panel stated in its first interim report, "the address list may be the most important factor in determining the overall accuracy of a decennial census" (National Research Council, 2000a:35).

The "three-legged stool" strategy outlined by the Census Bureau for the 2010 census calls for modernization of the Bureau's primary geographic resources:

- the Master Address File (MAF), the source of addresses not only for the decennial census but also for the Census Bureau's numerous survey programs; and
- the Topologically Integrated Geographic Encoding and Referencing System (TIGER), a database describing the myriad geographic boundaries that partition the United States.

The specific set of activities that the Census Bureau has described to achieve this modernization is known as the MAF/TIGER Enhancements Program (MTEP), an "8-year, roughly $500 million undertaking" (U.S. Department of Commerce, Office of Inspector General, 2003).

Given its nominal goal, the MTEP may be of paramount importance in terms of its potential impact on the quality of the 2010 census. However, the critical word in that statement is "nominal" since the term "MAF/TIGER Enhancements Program" suggests significant enhancements to both the MAF and TIGER. We do not argue that TIGER is unimportant; it is a critical geographic resource for census needs and it is in dire need of modernization. However, the MAF/TIGER Enhancements Program is oriented overwhelmingly toward TIGER and does little to enhance—to improve—the MAF. The Census Bureau's strategy for dealing with the MAF is, to an unfortunate degree, little more than routine maintenance—seemingly deferring active attention to the MAF until a complete block canvass very late in the census cycle (thus repeating a costly operation from 2000 that had been implemented as an eleventh-hour fix). The panel's unease regarding the Bureau's prospects for making material progress in improving its geographic resources for 2010 is further heightened by the apparent lack of comprehensive and realistic plans and schedules for the TIGER modernization effort.

In this chapter, we briefly review the development of both the MAF and TIGER (Section 3–A) before discussing the details of the MAF/TIGER Enhancements Program (3–B). Our general assessment of the program follows (3–C), with our particular call for attention to MAF improvement discussed separately in Section 3–D. Our recommendations—including designation of a MAF coordinator, strengthened geographic partnerships, and empirical justification of potential address sources—are detailed in Section 3–E.

3–A DEVELOPMENT AND CURRENT STATE OF THE MAF AND TIGER

Before we discuss the specific enhancements program that has been initiated by the Census Bureau, it is useful to first briefly review the nature and status of the two geographic systems addressed by the package.

3–A.1 The Master Address File

Purpose and Scope

The Master Address File (MAF) is the Census Bureau's complete inventory of known living quarters and business addresses in the United States and its island areas. The MAF contains a mailing address for each of those living quarters, if one exists. For housing units or living quarters without mail addresses, descriptive addresses (e.g., "2-story colonial with dormer windows") may be coded.

The MAF also includes an intricate set of flags and indicators that denote the operations that added or edited each address. It does not, however, record the date or time when an address was entered in the file or when it was modified. In principle, the MAF is a constantly evolving and continually updated resource; the "snapshot" of the MAF that is extracted and used to conduct the census is called the Decennial Master Address File, or DMAF.

Construction of the 2000 Census Master Address File

The concept of a continuously maintained MAF is a relatively new one; in the 1990 and earlier censuses, address lists were compiled from multiple sources prior to the census (e.g., lists were purchased from commercial vendors) and were not retained after the census was complete. The practice of maintaining the address list—to support not only the decennial census but also the Census Bureau's other survey programs—was initiated after the 1990 census. In part, writes Nash (2000:1), "a major impetus for this change was the undercounts experienced in the 1990 and earlier decennial censuses, nearly a third of which was attributed to entirely missing housing units." An initial MAF was constructed using the city-style addresses[1] on the Address Control File (ACF) developed for the 1990 census (Hirschfeld, 2000).

To populate the MAF, the Census Bureau "devised a strategy of redundancy using a variety of sources for addresses," thus "[assuming] responsibility for developing a comprehensive, unduplicated file of addresses" (Nash, 2000:1). Most prominent of the update sources were two that were endorsed by one of our predecessor Committee on National Statistics (CNSTAT) panels on the decennial census (National Research Council, 1995:5), which recommended that the Census Bureau "develop cooperative arrangements with states and local governments to develop an improved master address file" and that the U.S. Postal Service be given "an expanded role" in census address list operations. Both these recommendations were significant in that they required legislative authority in order to operate within the prohibition on release of confidential data codified in U.S. Code Title 13, the legal authority for census operations.[2] Congress granted this authority in the Census Address List Improvement Act of 1994 (Public Law 103-430).

The Delivery Sequence File One provision of the Census Address List Improvement Act authorized the Census Bureau to enter into a data-sharing arrangement with the U.S. Postal Service,

---

[1]A city-style address is one that can be specified by a numeric identifier (e.g., 305) in combination with a street name (e.g., Park Avenue), possibly with a specific subunit or apartment identifier. By comparison, non-city-style addresses are those that cannot be mapped to particular streets in this fashion, such as "Rural Route, Box 7" or a post office box.

[2]In *Baldrige v. Shapiro*, 455 U.S. 345 (1982), the U.S. Supreme Court ruled that the Census Bureau's "address list ... is part of the raw census data intended by Congress to be protected" under the confidentiality provisions of Title 13. Accordingly, the Court concluded that the Bureau's address list is not subject to disclosure under the Freedom of Information Act or under the discovery process in civil court proceedings.

under which the Postal Service would regularly share its Delivery Sequence File (DSF) with the Census Bureau.[3] The DSF is the Postal Service's master list of all delivery addresses served by postal carriers.[4] The name of the file derives from the Postal Service-specific data coded for each record along with a standardized address and ZIP code: namely, codes that indicate how the address is served by mail delivery (e.g., carrier route and the sequential order in which the address is serviced on that route). The DSF record for a particular address also includes a code for delivery type that is meant to indicate whether the address is business or residential.

Because the census is conducted largely through mailed questionnaires—most of which are subsequently mailed back—the U.S. Postal Service is a crucially important conduit in the census process. Moreover, the Postal Service is a constant presence in the field, servicing existing and emerging routes on a daily basis. For these reasons, securing access to the DSF was a major accomplishment. But while the DSF is an undoubtedly vital source of address information, it is incomplete for census purposes both because the list of mail delivery addresses is only a subset of the complete list of housing units in the United States and because it does not always properly distinguish multiple housing units within the same structure.

The Postal Service began sharing the DSF with the Census Bureau in the mid-1990s. Currently, as part of the Bureau's ongoing Geographic Base Support Program, new versions of the

---

[3]Specifically, the legislation text indicates that "the Postal Service shall provide to the Secretary of Commerce for use by the Bureau of the Census such address information, address-related information, and point of postal delivery information, including postal delivery codes, as may be determined by the Secretary to be appropriate for any census or survey being conducted by the Bureau of the Census. The provision of such information under this subsection shall be in accordance with such mutually agreeable terms and conditions, including reimbursability, as the Postal Service and the Secretary of Commerce shall deem appropriate."

[4]The list does not include general delivery addresses. Additional information on the DSF and commercial programs under which private companies are able to match their own address lists against the DSF can be found on the U.S. Postal Service Web site at http://www.usps.com/ncsc/addressservices/addressqualityservices/deliverysequence.htm [3/1/04].

DSF are shared with the Bureau twice per year and updates or "refreshes" to the MAF are made at those times.

Local Update of Census Addresses The Census Address List Improvement Act of 1994 also authorized the secretary of commerce and the Census Bureau to "provide officials who are designated as census liaisons by a local unit of general purpose government with access to census address information for the purpose of verifying the accuracy of the address information of the bureau for census and survey purposes." The act obligated the Census Bureau to "respond to each recommendation made by a census liaison concerning the accuracy of address information, including the determination (and reasons therefor) of the bureau regarding each such recommendation." The act thus permitted the Census Bureau to share with a local or tribal government for review and update the address data it had on file for that locality.

To preserve Title 13 confidentiality, the information to be disclosed to any particular locality was limited to address information and to the set of addresses for that area. Ultimately, the address information would be shared with local or tribal governments only if they signed an agreement to keep it confidential and to dispose of it when finished with review.

In August 1996, the Census Bureau initiated a program to acquire address list information from local governments. The Program for Address List Supplementation (PALS) contacted local and tribal governments (along with regional planning agencies) and solicited whatever lists of city-style addresses they maintained for their jurisdictions. However, the Bureau quickly concluded that the program was troubled: local address lists were not necessarily in computer-readable format, or were not formatted in such a way (including apartment and unit designators) as to match with the emerging coding system for the MAF. More significantly, response by local governments to an open-ended query for local address lists—ideally coded to the appropriate census block—was low. The program was officially terminated in September 1997 (U.S. Census Bureau, Geography Division, 1999).

The Census Bureau's next attempt at local geographic partnerships followed more closely the Address List Improvement Act by releasing parts of the Census Bureau's MAF for review rather than requesting entire address lists. The resulting program became known as the Local Update of Census Addresses (LUCA), though it is also occasionally referred to as the Address List Review Program. LUCA was conducted in two waves:

- *LUCA 98.* In 1998, local and tribal governments in areas with predominantly city-style addresses were given the opportunity to review the Census Bureau's address list. Bureau cartographers used blue lines to distinguish city-style from non-city-style address areas on the maps that defined eligibility for LUCA. As a result, LUCA 98 was said to target localities lying "inside the blue line."
- *LUCA 99.* In 1999, attention turned to areas outside the "blue line," those with non-city-style addresses.[5] Local and tribal governments were again invited to review Census Bureau materials, but this time the offer was to review block-level counts of housing units rather than actual addresses.

To participate in LUCA, local and tribal governments were required to identify liaisons who would handle the address list materials and take an oath of confidentiality. Materials were then sent to the governments, which had a specified time period to review them and submit any proposed changes. These changes were then reviewed by the Census Bureau, which often opted to reject part or all of the localities' suggested additions or deletions to the address list. An appeals process was set up under the auspices of the Office of Management and Budget (OMB), giving local and tribal governments a final opportunity if they found grounds to quarrel with the Census Bureau's judgments.

The Working Group on LUCA commissioned jointly by this panel and the Panel to Review the 2000 Census conducted an ex-

[5]The "blue line" designating LUCA 98 and 99 areas was not constrained to follow borders of whole geographic locations, so many places and counties were eligible to participate in both waves of LUCA. In some localities, the blue line did not cleanly distinguish between city-style and non-city-style areas, causing frustration for some LUCA participants (Working Group on LUCA, 2001). The process for delineating city-style-address areas should be refined for future LUCA-type programs.

Box 3.1 Results of LUCA Working Group Study

The Working Group on LUCA commissioned jointly by this panel and the Panel to Review the 2000 Census was composed of state and local government personnel who had been involved in their area's participation in the Local Update of Census Addresses (LUCA) program. The working group conducted a sample survey of LUCA-participant governments, inquiring about the techniques and resources they employed in order to complete a review of their local MAF segment. The working group report also provides detailed case study reports of LUCA participation, ranging in scope from rural communties to efforts at the state level to coordinate localities' participation in LUCA. The working group also analyzed available data on local and tribal government participation, including the numbers of addresses submitted by governments and accepted or rejected by the Bureau. However, available data did not allow for assessment of the number of completed census enumerations obtained using addresses added uniquely or in part by LUCA. The working group issued its final report in 2001.

The working group's analysis (Working Group on LUCA, 2001) led it to identify three principal barriers to effective local government particpation in LUCA:

- *Inaccurate designation of the "blue line" separating city-style and non-city-style address areas:* Localities expressed frustration at inaccuracies in drawing the boundaries used to distinguish the LUCA 98 program (reviewing specific addresses for city-style addresses) and the LUCA 99 program (reviewing block-level counts of housing units for non-city-style address areas). In some cases, the distinguishing "blue line" put portions of cities with seemingly valid city-style addresses into LUCA 99 territory, thus hampering localities' opportunity to offer specific address corrections.
- *Inconsistent designation of subunit identifiers:* Differences in reporting identifiers such as apartment or unit number made it difficult to match MAF extracts to local records.
- *Addresses rejected due to inaccuracies in TIGER:* As a consequence of out-of-date line features in the TIGER geographic databases, local address submissions were sometimes rejected because the addresses could not be geocoded based on current TIGER files. That is, seemingly accurate addresses were rejected because TIGER did not contain a new road or because the road's name or designation had changed.

The timing of the LUCA program leading to the 2000 census was also a concern to participants. Even large local governments with complete local geographic information files found it difficult to meet the turnaround time required for submission of addresses to the Census Bureau. The problem may have been compounded for local governments with less-developed geographic resources and in cases where manual review of address lists was the best or only available option; indeed, tight timelines combined with the requisite investment of resources may have dissuaded some governments from participation.

Box 3.1 (continued)

The working group found some evidence of increased participation and local cooperation in cases where state, regional, or county organizations worked to coordinate responses by multiple governments, sometimes providing a valuable "LUCA education" function. Improved training and guidance on the expectations of the program were identified as possible factors for increasing partipation in a LUCA-style program for the 2010 census.

tensive review of the LUCA process from the participants' (local government) perspective (Working Group on LUCA, 2001). The working group's principal findings are summarized in Box 3.1.

Block Canvass In the 1990 and earlier censuses, when address lists were not maintained from census to census but rather assembled before the decennial enumeration, a complete field canvass of the city-style addresses in designated mailout/mailback areas was a standard—but costly—operation. The Census Bureau had hoped to avoid a complete block canvass before the 2000 census; in introducing the Address List Improvement Act of 1994, U.S. Representative Thomas Sawyer expressed hope that "collection and verification of address information in primarily electronic format" from the Postal Service and local governments "will greatly reduce the amount of precensus field canvassing," an activity that he indicated had proven "expensive and often inaccurate."[6] Rather than a complete block canvass, the Census Bureau planned to target specific areas with coverage gaps and focus field canvass activities on those areas.

In spring and summer 1997, as a continuous MAF began to take shape, optimism about the completeness of DSF updates gave way to doubts when it also became clear that PALS was not proving an effective means to obtain address information from local and tribal governments. Internal evaluations convinced the Bureau that relying on DSF and LUCA alone could leave gaps in MAF coverage; in particular, the Bureau was concerned that "the DSF file missed too many addresses for new construction and

[6]Representative Sawyer's remarks can be found in the *Congressional Record* for the 103rd Congress, page H10618 (October 3, 1994).

was not updated at the same rate across all areas of the country" (National Research Council, 1999:39).

Accordingly, the Census Bureau opted to change course and conduct a full canvass of addresses in mailout/mailback areas "in a manner similar to the traditional, blanket canvassing operations used in prior censuses." The Bureau noted that the change would incur a large expense, but, recognizing the Bureau's concerns, a previous CNSTAT panel "strongly endorse[d] this change in plans" (National Research Council, 1999:25,39).

Plans for the complete block canvass overlapped with the emerging plans for the LUCA program. The Bureau originally planned for LUCA 98 to obtain feedback in early 1998, so that resulting changes to the MAF would be ready for the block canvass in late 1999. However, delivery of MAF segments to most participating LUCA 98 localities was delayed. This led to a revised plan that LUCA 98 changes would be compared to the MAF after block canvassing was complete. Further delays led to abandonment of a reconciliation operation in which discrepancies between LUCA and block canvass observations would have been reviewed with localities; instead, localities received a list of accepted and rejected addresses in LUCA's "final determination" phase and were given 30 days to submit appeals to OMB's address list appeals office (Working Group on LUCA, 2001).

3–A.2 The TIGER Database

Purpose and Scope

The TIGER database is, effectively, a cartographic resource that defines a complete digital map of the United States and its territories. It is intended to capture not only visible features—the centerlines of streets, rivers, and railroads, and the outlines of lakes, for instance—but the myriad political and administrative boundaries that may not correspond exactly with visible physical locales. Accordingly, the TIGER database includes the political geography of 3,232 counties or county-level equivalents, more than 30,000 county subdivisions or minor civil divisions, and more than 20,000 named places, among other political units.

Of the many geography types defined by the TIGER database, the most important are the boundaries of census blocks. Census blocks are the smallest unit of geography for which basic population data are tabulated in the census, and these block-level data are aggregated to form political and other administrative boundaries. TIGER's primary function in census operations is *geocoding,* the matching of a given address or location to the census block in which it lies. Once a location has been matched to the correct census block, its location in higher-level geographic aggregates constructed from blocks is also known, and so census returns may be properly tabulated by geographic unit.

In addition to the geocoding function, the Census Bureau has relied on TIGER for three other major uses (O'Grady and Godwin, 2000; U.S. Census Bureau, 2001b):

- *geographic structure and relational analysis:* the definition of how one geographic area relates to another, which is important for being able to aggregate small units like blocks into coherent higher-level geographic entities;
- *geographic definitions:* a repository for the current definitions of geography levels recognized by the Bureau; and
- *map production:* the basis for printed maps used by census enumerators, and other geographic products.

The Census Bureau's full TIGER database contains both point and line features; in particular, points define the location of known housing units in areas without city-style addresses. However, most public exposure to TIGER comes via TIGER/Line files, a public excerpt of the TIGER database that contains only linear features such as roads, rails, and political boundaries (not specific housing unit locations). The TIGER/Line files, which contain complete street coverages with address ranges, helped facilitate the emergence and growth of the geographic information systems (GIS) industry.

The TIGER database is one part of a larger TIGER system, which includes the support structure of hardware and software necessary for maintaining the database. TIGER was initially created using a unique, home-grown language developed by the Census Bureau, and various software programs to update the database and to produce maps were similarly written to accommodate this customized internal language. As we will discuss, the proposed MAF/TIGER enhancements make changes in both the database and system senses, improving the content of the database as well as overhauling its support machinery.

How the TIGER Database Began

The TIGER database was developed by the Census Bureau, with assistance from the U.S. Geological Survey (USGS), to support the 1990 census. "TIGER began life as a patchwork quilt of data sources" (O'Grady and Godwin, 2000:6), two of which were primary. One of these sources was the Geographic Base File/Dual Independent Map Encoding (GBF/DIME) files used by the Census Bureau to do address matching to street segments in the 1980 census. The GBF/DIME files foreshadowed TIGER in that they applied topological principles in piecing together points, lines, and polygons (Hirschfeld, 2000); they also began the move toward including more than streets and roads in census maps, adding features such as water, rail, and invisible boundaries. But the files were limited in scope, covering only the urban centers of 276 metropolitan areas—"less than 2 percent of the land area but 60 percent of the people in the United States" (Carbaugh and Marx, 1990). To complete the geographic coverage of the nation, the address reference information in the GBF/DIME files was merged with computer-coded versions of the water and transportation features defined by the USGS series of 1:100,000-scale topographic maps (Marx, 1986).

As O'Grady and Godwin (2000) note, "accuracy was crucial" when TIGER was first assembled "but only in a relational sense." "The coordinate information presented in the TIGER/Line files is provided for statistical analysis purposes only," wrote Carbaugh and Marx (1990); "it is only a graphic representation of ground truth." Put another way, the priority in early TIGER was to achieve basic functionality for census purposes, which meant favoring relational accuracy (describing how geographic features relate to each other, such as whether census blocks are adjacent) over positional or locational accuracy (precise location of geographic features relative to a chosen standard). O'Grady and

Godwin (2000:5–6) recall that the Census Bureau drew on properties of the USGS maps in publishing the following positional accuracy statement in the documentation for TIGER/Line files released in 1995:

> The positional accuracy varies with the source materials used, but at best meets the established National Map Accuracy standards (approximately ±167 feet) where 1:100,000-scale maps from the USGS are the source. The Census Bureau cannot specify the accuracy of feature updates added by its field staff or of features derived from the GBF/DIME-Files or other map sources. Thus, the level of positional accuracy in the 1995 TIGER/Line files is not suitable for high-precision measurement applications such as engineering problems, property transfers, or other uses that might require highly accurate measurements of the [Earth's] surface.

The overall positional accuracy of early TIGER was also limited by shortcomings in the GBF/DIME files, which were also oriented toward relational accuracy. In particular, Census Bureau enumerators and staff later found that "hydrographic features are not represented well" in TIGER database segments derived from the GBF/DIME files (Rosenson, 2001:1).

Updates to TIGER

During the 1990s, the TIGER database was updated using additional sources, each with unique (and often unknown) levels of positional accuracy. Among those sources are the following programs that are likely to continue during and after the MAF/TIGER Enhancements Program, although exactly how and when the resulting information will be incorporated—and how the programs might be restructured—is as yet unspecified:

- *Boundary and Annexation Survey (BAS):* an ongoing voluntary survey in which TIGER-generated boundary maps are sent to local and tribal governments for review and update.
- *MAF Geocoding Office Resolution (MAFGOR):* a program in which city-style address records from the Postal Service Delivery Sequence File (DSF) that cannot be geocoded in TIGER are referred to census regional offices for review.

- *Targeted Map Update (TMU):* a regular program in which census field staff update address ranges, add new streets, and update feature names in selected areas.
- *Digital Exchange (DEX):* a system that draws on local and tribal geographic database files.

Of these, the DEX system (Rosenson, 2001) developed in the late 1990s is of particular interest as improvements to its capabilities will be a major part of TIGER realignment. DEX does not directly manipulate local and tribal geographic files but rather a processed extract known as an "exchange file." The system is strictly limited to working with road features and the attributes associated with them, including ZIP codes. The exchange file derived from a local geographic file is a street centerline database coded in TIGER format. This exchange file is then matched to the TIGER file based on both spatial location and attribute information (e.g., street name), beginning with matches on the intersection points between named road features in each file.

After matching, one of the files is "rubber-sheeted"—meaning that its features are adjusted to better match attributes in the other file, with neighboring attributes being adjusted simultaneously, as necessary. As Rosenson (2001) notes, this "rubber-sheeting" can be done to either file but, at least in early DEX implementation, the process could introduce topological errors such as lines that cross each other without a system-defined point marking their intersection. Thus, in order to preserve TIGER's topological structure, DEX manipulates the local "exchange file" to match certain TIGER features.

Though some DEX capability had been developed and selected local geographic files were obtained prior to the 2000 census, active TIGER updating using DEX was deferred during the actual conduct of the 2000 census.

The Need to Modernize

The development of TIGER is a milestone of which the Census Bureau should be extremely proud. A homegrown database management system constructed to manipulate an enormously complex network of visible and invisible boundaries, TIGER became an exemplar of what a GIS can do. The example of TIGER—and, significantly, the public availability of TIGER/Line files, a full and fine-scale public atlas of the United States—touched off a commercial GIS revolution. Businesses and organizations of all sizes are continuing to learn the power of spatial data analysis, and the work of TIGER to bring together and make publicly available base geographic layers helped make that possible. TIGER successfully satisfied the operational demands of two decennial censuses. The coding system may be (in computer years) old and the structures arcane, but it is a rare in-house software product that can successfully cope with a production cycle of billions of printed maps and millions of addresses for geocoding as TIGER did in the 1990 and 2000 censuses.

But, as is the case with some revolutions, the first entrant ushers in tremendous change and then is unable to keep pace with the new world thus created; so it is with TIGER. Though the text-based TIGER/Line files are parsable by commercial GIS applications, the native TIGER database structure is not compatible with modern database tools. As a result, it has not been possible to directly update TIGER's street coverages using the GIS files updated and maintained by local and tribal governments. The Census Bureau's unique role in delineating census blocks—the base units that are aggregated to form most political districts—and ongoing programs such as the Boundary and Annexation Survey (BAS) give the Census Bureau advantages in defining the invisible political boundaries that cross-cut the nation. But commercial GIS has made it possible for external companies and local and tribal governments to build on the TIGER/Line base, realigning features when errors are found and making updates to street, rail, water, and other features to a degree that Census Bureau resources have not permitted.

3–B THE MAF/TIGER ENHANCEMENTS PROGRAM

The Bureau has set forth five objectives as essential steps in a comprehensive MAF/TIGER modernization:

1. improve address/street location accuracy and implement automated change detection;

2. implement a modern processing environment;
3. expand and encourage geographic partnership options;
4. launch the Community Address Updating System (CAUS), which has also been known as the American Community Survey Coverage Program; and
5. implement periodic evaluation activities and expand quality metrics.

They are spelled out with subtasks in the following sections.

3–B.1 Objective One: Address/Street Location Accuracy

Objective One—the actual realignment of TIGER geographic features—is the centerpiece of the MTEP, enough so that it has acquired an acronym of its own. The contract to carry out Objective One—also known as the MAF/TIGER Accuracy Improvement Project (MTAIP)—was awarded to the Harris Corporation of Melbourne, Florida, in June 2002.

As described in documentation provided to the panel, the basic subtasks envisioned under Objective One are as follows:

1. correct (in TIGER) the location of every street and other map feature used by field staff and governmental partners for orientation, as well as the location of every boundary used for tabulation of decennial census and household survey data;
2. correct (in the MAF) the location of every housing unit and group quarters from which the decennial census and the household surveys collect data; and
3. implement an effective change detection methodology to document the location of every new street and living quarters, along with the street name and address for each.

Means of Updating Accuracy

As it has been explained to the panel, the basic idea of Objective One is to perform a single, extensive update of TIGER for each county based on an external source with, presumably, more current and accurately positioned feature information. These outside sources may include GIS files developed and maintained by local or tribal governments, commercial GIS files, or digital orthophotography/aerial photography. Once the TIGER data for a county are realigned, they can be continually updated through change detection—for instance, features may be added as a result of comparison of TIGER to newer aerial photographs of a region. Through this strategy—extensive initial realignment, followed by change detection—the Census Bureau hopes to maintain TIGER so that its features are current to within one year.

This general framework provides great flexibility for the Census Bureau and its contractor to implement the TIGER update; at present, however, to the extent that plans have been shared with the panel, this flexibility translates into little specificity.

The Census Bureau has established a cartographic accuracy standard for the realigned TIGER database: 7.6 meters CE95, meaning that, for a sample of control points measured on the ground and the corresponding locations in the geographic database, at least 95 percent of the database-recorded points should lie within a 7.6-meter radius of the corresponding ground-recorded points. According to the Census Bureau's presentation to another National Research Council committee, the 7.6 meters CE95 resolution was chosen because it is the minimum required accuracy "to support use of GPS equipped handheld computers to achieve 99.6% geocoding accuracy for tabulations"; it is also said to be "based on accuracy of enumerator's GPS-equipped hand-held computer and relationship of enumerator to street centerline" (LaMacchia, 2003).

The Census Bureau informed the panel in September 2003 that it expects 1,200 of 3,232 total counties to be able to provide local files meeting this accuracy standard (Jackson, 2003:2). The request for proposals (RFP) issued to solicit contractor bids to perform Objective One indicates the Census Bureau's strong preference to use local or tribal government GIS files as the update source whenever possible.[7] But, based on the information

[7]The RFP and other documents related to Objective One, the MAF/TIGER Accuracy Improvement project, are archived at http://www.census.gov/geo/mod/maftiger.html [3/1/04].

known to the panel, no approach has been specified for the balance of counties for which local GIS files are not available or do not meet the Bureau's accuracy standard. The Census Bureau has conducted experiments using subcontractors to perform updates based on digital orthophotographs and other image sources. Other potential means of collecting the geographic information include buying commercially available GIS files or using field staff to collect GPS trace data while driving or walking streets. It is as yet unclear which of these (or other) mechanisms the Census Bureau and the Harris Corporation will favor in the absence of local files (or when local files are of insufficient quality) to perform the initial, global realignment. In the omnibus appropriations bill for fiscal 2004, House and Senate appropriators "[direct] the Secretary of Commerce to take all necessary measures to reduce the payment for information currently available from certain governments" and "to utilize global positioning system technology and aerial photography to update existing information only if these measures are shown to be cost effective" (H. Rept. 108-141, citing H. Rept. 108-221).

As it is unclear what exact source will be used for the initial realignment in particular counties, it is even less clear what source will be used to update TIGER files in the change detection process, and with what frequency this will be done.

Priorities

Franz (2002) described the following priority structure that the Census Bureau has identified for carrying out Objective One realignment, with the first being the top priority:

1. linear feature realignment across all areas;
2. establishing/correcting structure locations in areas outside the 2000 census mailout/mailback area;
3. establishing/correcting structure locations inside the 2000 census mailout/mailback area; and
4. establishing/correcting locations for residential structures over nonresidential structures, in carrying out the previous two steps.

Schedule

Under plans developed in 2002, the Census Bureau and the Harris Corporation are supposed to realign counties on the following timetable: 250 in fiscal year 2003; 600 in 2004; 700 in 2005;[8] 700 in 2006; 600 in 2007; and 382 in 2008. In principle, change detection to make further alterations is supposed to begin when counties are complete, so that 250 counties are slated for change detection in fiscal year 2004, 850 in fiscal 2005, and so forth, until all counties are handled using change detection methods in 2009.

3–B.2 Objective Two: Modern Processing Environment

Objective Two of the Enhancements Program targets TIGER in the systems sense, modernizing the structure of the database. The current homegrown TIGER system suffers from key limitations, prominent among them the inability to directly link with commercial GIS packages (and hence local and tribal GIS files maintained using those packages) and the limitation that only one module (county) of TIGER may be "checked out" for updating at a time. Changes to the database structure also require that the suite of support software used to generate products from TIGER—for instance, to print maps for field enumeration—must be reauthored and tested.

The Census Bureau's stated subtasks for Objective Two are as follows:

1. make maximum possible use of commercial off-the-shelf (COTS) and geographic information systems (GIS) tools to allow for rapid development of new applications; and
2. customize the COTS/GIS tools to the minimum extent possible to avoid schedule and cost obstacles when the COTS/GIS vendors deploy new versions of their software.

Under original timelines specified by the Census Bureau, fiscal 2003 was to be the peak year of Objective Two work, with

[8]The figure of 600 counties was included in the detailed description of the Bush administration's budget request to Congress for fiscal year 2004; likewise, the administration's budget request for fiscal 2005 confirms the goal of realignment of 700 counties in fiscal 2005.

some slight drop-off in fiscal 2004. Residual effort was expected in fiscal 2005 and 2006, with Objective Two not listed as an activity in 2007 or later years.

We discuss the tasks to be accomplished under Objective Two in greater detail in Section 6–C.

3–B.3 Objective Three: Geographic Partnerships

Objective Three acknowledges the crucial role of state, local, and tribal governments in maintaining geographic resources, not only for the TIGER realignment of Objective One but for continued update of the MAF, as in the LUCA program.

Subtasks of Objective Three identified by the Census Bureau are as follows:

1. devise and deploy new strategies to communicate more effectively with governments to increase the level at which they participate in MAF/TIGER review and update activities;
2. devise and deploy new ways to integrate more effectively the address list review, street update, and boundary reporting activities that now exist as separate programs; and
3. establish new partnerships with other federal agencies and private-sector firms that have GIS and address files with information of value to an accurate and complete MAF/TIGER.

Under original timelines shared with the panel, fiscal 2004 was scheduled to be the peak year of Objective Three work. The levels of effort expected on this objective in each of the years 2003 and 2005 through 2010 are to be roughly equivalent.

3–B.4 Objective Four: Community Address Updating System

Briefly known as the ACS Coverage Program, the Community Address Updating System (CAUS) is the address list update component of the proposed American Community Survey (ACS). The basic idea of the program is to make use of the continued field presence that would be necessary to conduct the ACS, allowing ACS enumerators the opportunity to provide geographic updates. One hope is that the ACS enumerators might be particularly helpful in identifying geographic and housing changes in rural areas, where local and tribal files might be less detailed (or unavailable).

The Census Bureau has identified the following subtasks for Objective Four:

1. focus predominantly on rural areas, in which the Census Bureau has concluded that the U.S. Postal Service's Delivery Sequence File (DSF) does not effectively identify the existence or location of new housing units; and
2. provide address list (and street) updates beyond what can be identified through the current twice-yearly DSF "refresh" process to ensure a uniformly accurate sampling frame nationwide for the ACS and the other household surveys.

Through contractors, the Census Bureau has developed prototype Automated Listing and Mapping Instrument (ALMI) software, making use of a GPS receiver and a laptop computer. The ALMI system could permit ACS enumerators who encounter a new street that is undefined in TIGER to record a GPS trace as they drive along the street and to note the location of houses along that street; these inputs could later be converted to TIGER.

The anticipated level of effort that the Census Bureau expects to expend on Objective Four is roughly equivalent during each of the fiscal years 2003–2010.

3–B.5 Objective Five: Evaluation and Quality Metrics

Finally, Objective Five concerns the assessment of progress and quality; subtasks identified by the Census Bureau for this Objective include the following:

1. provide quality metrics information that will guide (target) areas in need of corrective action beyond the changes identified in the change detection and CAUS activities;
2. document progress toward improving the accuracy and completeness of the street, address, and boundary information in MAF/TIGER; and

3. ensure the availability of accurate and comprehensive metadata that meet federal standards for the information in MAF/TIGER.

The anticipated level of effort that the Census Bureau expects to expend on Objective Five is roughly equivalent during each of the fiscal years 2003–2010.

3–B.6 Update on Enhancements Program Progress

The Census Bureau's goal for fiscal 2003 was to complete Objective One TIGER realignment for 250 counties. At the panel's September 2003 meeting, the Census Bureau reported that it was set to meet that goal, with 244 counties already completed. Only 60 of those completed were realigned by the Harris Corporation, which holds the Objective One contract and is responsible for realigning TIGER data for 600 counties in fiscal 2004; the remainder represent work from other contractors on earlier pilot projects. Eight of the 60 files were said to have been returned to Harris for "rework" because of unspecified problems (Jackson, 2003:3).

As of September 2003, the Census Bureau had collected 1,038 GIS files from local and tribal governments and was testing them to see whether they met the Bureau-imposed 7.6 meters CE95 accuracy standard. In September, the Bureau reported that it had collected ground control points with GPS receivers for 777 of the files; results showed an equal divide, with 390 meeting or exceeding the 7.6-meter standard and 387 failing (Jackson, 2003:3). In October, the Bureau submitted an update to the panel, now stating that 826 files had been tested, with 461 of these meeting the standard and 365 failing (U.S. Census Bureau, 2003e:1). Noting that many of the 365 subuniform-standard files nonetheless appeared to be more positionally accurate "in the densely settled extent of their coverage," the Bureau and the Harris Corporation are said to be developing "a method for utilizing the accurate sub-extent of local GIS files (with Harris supplying and utilizing an accurate source for the balance area) by the end of fiscal year 2004" (Jackson, 2003:3).

Objective One of the Enhancements Program faces a heavier workload in fiscal 2004, with the goal of realigning 600 counties. The Bureau expects that the Harris Corporation will use local or tribal GIS files to update 350 of those counties and that, "for the remaining 250 counties, Harris will acquire, evaluate, and use sources such as commercial GIS files, imagery, and field-collected GPS road centerline data" (U.S. Census Bureau, 2003e:2). As we discuss in Section 3–C.1, the Bureau has provided no indication as to which counties will be targeted for update in 2004.

After TIGER files have gone through initial realignment, they are then supposed to be subject to updating using change detection—that is, using a newer-vintage local GIS file or aerial photography to automatically find new streets or structures. According to the Bureau, "requirements and methodology for detecting change (growth) for areas that have been realigned" are to be drawn up in fiscal 2004 (Jackson, 2003:4). To what extent, if at all, delays in finalizing these requirements result in delays in updating the 250 2003-realigned or 600 2004-realigned files remains to be seen.

We will discuss the status of Objective Two, the database redesign and conversion, in Section 6–C. Objectives Three and Four (partnerships and CAUS, respectively) remain largely unplanned; a "program master plan for geographic partnerships" is slated to be developed during fiscal 2004 and CAUS implementation (like the ACS) was postponed due to late closure on the fiscal 2003 budget.

3–C ASSESSMENT OF GEOGRAPHIC MODERNIZATION EFFORTS

3–C.1 Locational Accuracy of TIGER

Problems with the positional accuracy of TIGER have been apparent to the Census Bureau and its users for some time; anecdotal experiences of problems with TIGER representations were reported by field enumerators during the 2000 census and in feedback from local and tribal governments that participated in LUCA (Working Group on LUCA, 2001). Quantitative evidence of TIGER discrepancies can be found in Liadis (2000), the report of a Census Bureau experiment that collected GPS readings for approximately 6,700 "anchor points" spread across selected census tracts in eight counties. Distances were computed between these "ground truth" coordinates and the longitude/latitude combination coded in TIGER. The results show evidence of considerable local variation, even across tracts within the same county. The distance between TIGER representation and ground truth varied according to the method used to introduce the point into TIGER. Somewhat ironically, more recent update programs—which added features by digitally inserting them as freehand drawings—accounted for the largest deviations from ground truth, while pre-1990 sources (e.g., GBF/DIME files) and programs involving direct use of local and tribal geographic files (e.g., DEX) generally came closest to true locations. The Census Bureau's Geography Division also conducted pilot experiments comparing TIGER coordinates for small geographic samples to a combination of GPS coordinates and commercially available cartographic databases (U.S. Census Bureau, Geography Division, 2000) and to digital orthophotos giving an aerial view of ground features (O'Grady, 2000).

Though the full extent of TIGER inaccuracy may be unknown, there is enough evidence available that the panel endorses the aims of Objective One. Errors in the placement of roads, boundaries, and other geographic features are sufficiently serious and numerous that the TIGER database is in need of a comprehensive update. Moreover, raw TIGER/Line files cannot be fully trusted for routine GIS- and non-GIS-related tasks.

Given that locational error in TIGER is extensive enough to require correction, it follows naturally that accomplishing the basic task envisioned under Objective One is essential to the modernization of the census. GPS coordinates collected by PCDs are useful only to the extent that they can be accurately placed on base maps with streets and other key features. An accurately aligned TIGER, faithful to polygonal features such as municipal boundaries, can be passed along to localities and made available on the Internet, thereby allowing local and tribal entities the opportunity to report changes made to both linear (e.g., road and railroad) and polygonal features (e.g., administrative borders collected by the Boundary and Annexation Survey) in a more efficient and accurate way. If localities can readily utilize an aligned TIGER for geocoding their own address files, comparisons with (and updating of) the MAF may eventually become routine.

Hence, the panel supports Objective One of the Enhancements Program and is heartened by the general steps taken to accomplish the objective. In particular, the panel views the acquisition of an outside contractor as a sign of significant progress, rather than keeping the process of TIGER updating a purely in-house operation. As Census Bureau staff noted in an interview, it is indeed a "very major departure for us" to seek external help in retooling TIGER, but "we've come to the conclusion [that] we need to take advantage of [vendors'] expertise and understanding" (O'Hara and Caterinicchia, 2001).

In the panel's assessment, the Census Bureau deserves high grades for its determination to fix a major problem as well as for the boldness of the approach outlined in Objective One. That said, concerns about the work remain, and the plausibility of the Census Bureau's ambitious realignment timetable would be bolstered considerably through attention to the following:

- a detailed work plan, including the order in which counties will be initially updated;
- realistic estimates of the number of available state and local GIS files that meet, in part or in full, the Census Bureau's chosen positional accuracy standard for the realigned TIGER;
- a clear plan for the evaluation of initially realigned TIGER files in order to inform future realignment as well as to recalibrate the Objective One timeline and budget; and
- specification of plans for the postrealignment change detection program.

A point of some contention between the panel and the Census Bureau has been the order in which Objective One realignment will be performed. Aside from indicating that jurisdictions involved in mid-decade census tests or dress rehearsals will be given priority, the Census Bureau has not given a clearer idea of how it expects the flow of county-by-county processing to proceed. The notion of ordering is understandably somewhat sensitive, since no locality would relish being last in the queue. However, the ambitious timetable laid out earlier in this chapter is unrealistic—at best—without some sense of ordering. The alternative—effectively starting 3,232 independent updating efforts simultaneously and hoping that 850 fall into realignment by the end of 2004—does not inspire confidence. There is no right answer to the question of ordering—conceivable mechanisms include starting with urban counties or rural counties, starting with original GBF/DIME areas, sequencing by population, or sequencing by some assessment of how out of alignment TIGER is for an area. But providing some structure to the task seems essential for measuring progress toward complete realignment and could add plausibility to the hypothesized timetable.

At the panel's September 2003 meeting, the Census Bureau acknowledged this concern, noting that "it has been a challenge to balance the desire to establish a firm and detailed county-by-county schedule for the realignment effort on the one hand, and [maintain] the flexibility to take advantage of newly emerging tribal and local source data on the other hand." The Census Bureau now indicates that the listing of local source files to be realigned "will be firmed up quarterly, 30 days prior to the start of the quarter" (Jackson, 2003:3).

In addition, a subtle point raised in our earlier discussion of the Census Bureau's Digital Exchange (DEX) program deserves fuller explication. Given two GIS files (a local file and the TIGER data), a "rubber-sheeting" process manipulates certain matched features in one file to conform to the other, shifting related features automatically. The Census Bureau's early DEX system altered the local file to follow known features in TIGER in order to avoid topological bugs that might result otherwise—a justifiable choice, perhaps, but one that runs counter to the purpose of updating the presumably misaligned TIGER based on presumably accurate local files. We hope and trust that this approach has been rectified as the Bureau has developed procedures with its contractor; the Bureau noted in its update of Enhancements Program progress that "Harris is required to align TIGER road features exactly to the source data (which, again, must meet or exceed the 7.6 meter accuracy standard) as well as maintain TIGER's topological integrity" (U.S. Census Bureau, 2003e:2).

Further empirical information on discrepancies between local file content and existing TIGER topology (and their resolution), along with additional detail on how the Harris Corporation's alignment tools handle topological gaps and generally manage the conflation between local and TIGER files, could strengthen confidence in the finished product.

3–C.2 Balance of the MAF/TIGER Enhancements Program

The panel applauds the Census Bureau's efforts to adopt GPS technologies and a modern processing environment using COTS products to achieve Objectives One and Two. We comment on Objective Two—discussing major points of concern—in Section 6–C, in the context of the census technical architecture.

We also note that the Census Bureau has made some steps toward establishing metrics to evaluate improvements in accuracy, as called for by Objective Five. Work with contractors has brought about an image-based rough assessment system that allows accuracy checks on incoming files, as well as progress on evaluation of files on the basis of control points, and a soon-to-be-installed system for quantifying and tracking TIGER errors over time. It is essential, in our view, that quality assessment through such metrics be an ongoing and well-timed process so that updating of the database achieves the apparent goal: information in TIGER maintained to a currency of one year or less at all times.

As elaborated in Chapter 8, the panel emphatically believes that Objective Five is a crucial part of the Enhancements Program and should lead to the development of general quality metrics for all of the Census Bureau's geographic programs. However, with respect to progress on Objective Five, two comments must be made. First, it is possibly telling that neither the Census Bureau's presentation to the panel in September 2003 (Jackson, 2003) nor the subsequent update in October (U.S. Census Bureau, 2003e) addressed progress on Objective Five. Beyond the diagnostic function for local files to be included in Objective One realignment, general progress on metrics for TIGER quality and coverage is not known. Second, and more fundamentally, all discussion of Objective Five activities—see, for instance, the

Bureau-identified subtasks in Section 3–B.5—has focused almost exclusively on quality metrics for geographic coordinates, not for addresses. That is to say, to the extent that Objective Five is defined at present, it is focused on TIGER and TIGER realignment; it is not focused on the MAF, a fact that we believe is symptomatic of a larger lack of attention on the Bureau's part.

3–D WEAKNESS: ENHANCING THE MAF

It is clear that the MAF/TIGER Enhancements Program has the potential to enhance TIGER, making necessary improvements given known problems with TIGER accuracy. But, for the sake of census accuracy, a more important question is how the program will enhance the MAF—that is, how it will add new addresses, screen for duplicates, and generally ensure that address rosters are as complete and accurate as possible. On this score the Enhancements Program falls seriously short, in our view, due to the lack of development of Objectives Three and Four. More generally, the Census Bureau's current strategy shows relative inattention to MAF improvement and, worse, shows signs of repeating costly errors from the 2000 experience.

The magnitude of the Objective One task of realigning TIGER features—and the monetary cost associated with it—give the Enhancements Program a TIGER-centric feel. And Objectives One, Two, and Five seem to speak to the MAF largely as it inherits its quality from TIGER. Indeed, the Bush administration's budget messages to Congress for both fiscal 2004 and 2005 describe the geographic leg of the Census Bureau's 2010 strategy as a plan for "enhancing the Census Bureau's geographic database *and associated address list*" (emphasis added). In line with our comments in opening this chapter, the MAF is too critical to the quality of the census and other survey programs to be treated merely as an add-on.

3–D.1 Current Plans for MAF Updates for 2010

The Census Bureau argues that the combination of three activities—"the ongoing MAF/TIGER updating using the Delivery Sequence File, CAUS, and enhancements included in the proposed MAF/TIGER modernization initiative"—"should result in an up-to-date address list for the entire United States" (U.S. Census Bureau, 2003c:11). More specifically, the update strategy is based on a rough urban/rural dichotomy:

- The Postal Service's DSF is intended to be the address update source "in areas where DSF addresses can be assigned a physical location, such as urban areas with city-style addresses" (U.S. Census Bureau, 2003c:9).
- "In rural areas with non-city-style addresses," the Bureau argues that the DSF updating process "cannot be used," and so the Census Bureau intends to update this segment through CAUS. The Bureau indicates that the areas for which DSF updates cannot be used "encompass the majority of the Nation's land area and about 15 percent of the population" (U.S. Census Bureau, 2003c:9).

These update sources are to be supplemented in the MAF/TIGER Enhancements Program, which we interpret to mean a successor to the 2000 census LUCA program under Objective Three.

The backbone of the Census Bureau's update strategy is the twice-yearly "refresh" that comes from the Postal Service's Delivery Sequence File. These regular updates are considered to be part of the Bureau's Geographic Support Base Program, not the MAF/TIGER Enhancements Program. While the DSF is certainly an important source of address information, reliance on the DSF as the principal source of address updates for (by the Bureau's estimate) 85 percent of the household population raises at least two concerns:

- *Historical precedent in the 2000 MAF-building process.* As we indicated earlier, DSF updates were previously viewed by the Census Bureau as a primary address source after the 1994 passage of legislation that enabled sharing of this information with the Postal Service. However, the Bureau perceived problems with the level of DSF coverage in fast-growth and new construction areas and had to initiate a costly complete block canvass (National Research Council, 1999) in an attempt to ensure comprehensive coverage.

- *Limitation of DSF to mail delivery population.* Again, by definition, the DSF is intended to document mail delivery addresses, which is not equivalent to the complete list of housing units in the United States.

The Census Bureau's planned activity to update addresses in rural areas is CAUS, which—to briefly review—is an associated program of the American Community Survey (ACS). Under CAUS, ACS field representatives would list addresses (and update streets, using traces from a GPS receiver) through a laptop computer-based tool known as the Automated Listing and Mapping Instrument (ALMI). However, general concerns raised by dependence on CAUS as an address update source include the following:

- *Linkage to ACS funding.* Full and sustained funding for the ACS has not yet been secured; consequently, the budgetary viability of CAUS is not known. Implementation of CAUS must also await full mobilization of ACS support staff (which will presumably entail more time as well, as the establishment of ACS operations takes priority), which will add to the delay in the possible receipt of CAUS updates. Finally, the number of CAUS field personnel will be linked to the number of ACS enumerators. While it is hoped that budget commitments to ACS will not oscillate, the effectiveness of CAUS could be impaired if ACS funding is not stable over the years.
- *ACS workload management.* It is unclear how much time and manpower ACS managers will commit to the side work of the address listing given the ambitious timetable of ACS data collection.
- *Unclear/unspecified mechanism for targeting areas for update.* The plans for deployment of CAUS representatives to collect information in particular geographic areas are as yet unspecified. One approach might be for enumerators to list new streets or developments they find by happenstance in carrying out their regular ACS work, but that is surely an unreliable means of covering the entire rural population. The draft ACS operations plan indicates that "ACS planners

> [will] use various methods for identifying where coverage is insufficient," including "work with community officials to acquire information about new addresses, new streets, and/or areas of significant growth" (U.S. Census Bureau, 2003c:10). But, again, the mechanics of this targeting are uncertain.

The third element in the address update strategy—a LUCA-type program—is a topic we will discuss in greater detail in Section 3–E.5. But, for the purpose of the argument at hand, the major concern regarding a new local address review program is simply that no prototype plans have yet been developed.

3–D.2 Block Canvassing

In our second interim report, the panel commented (National Research Council, 2003a:66):

> We assume that the Bureau hopes to avoid a complete block canvass prior to the 2010 census, given the cost of that operation and that it was treated as a last resort in 2000.

Our supposition was that the Census Bureau would pursue targeted block canvassing—identifying selected geographic areas with sufficiently fast growth or other characteristics to warrant a thorough precensus address list check.

In responding to the interim report at our September 2003 meeting, the Census Bureau expressed surprise at this statement, maintaining that a full block canvass was always part of the Census Bureau's 2010 plan. We respectfully disagree; part of the tenor we recall in early discussion of the MAF/TIGER Enhancement Program was the need for continuous address updating over the next decade in order to avoid a block canvass. The Census Bureau's document on projected life-cycle costs of the 2010 census suggests the desire to replace a last-minute canvass with continuous updating. "While address building and TIGER updating occurred to a limited extent over the decade leading to Census 2000," the document says, "the major updating activities occurred during 1998–99 and involved expensive, complex, labor-intensive field operations." As a result of regular DSF updates and local and tribal updates, "the 2010 Census will be armed with a far more comprehensive, timely, and accurate address list—one of the best predictors of a successful census—without the complexity, risk, end-of-the-decade costs, and last minute address building costs" (U.S. Census Bureau, 2001a:3–4).

Regardless of when the idea reemerged, the panel acknowledges with some concern that a full block canvass now appears to be part of the Census Bureau's plan, though no detailed schedule or specifications are known to us, nor have any changes to the operation from its 2000 census implementation been described.

3–D.3 Conclusions

We understand the draft nature of the current 2010 planning documents and are thus hesitant to quote from them extensively. But the Census Bureau's comments on address list issues in its draft baseline design document (Angueira, 2003b:3) suggest an emerging direction that could potentially be so damaging to a quality census that they merit detailed examination. The comments begin:

> When address list updating gets underway in 2009, census geographers and field staff will be working with an address list unprecedented in its accuracy and completeness.

Were nothing to be done with the Master Address File between now and 2010, the statement would hold by virtue of the fact that (unlike censuses before 1990) the 2000 MAF was not discarded following the census. Just as the 1990 Address Control File was the lead contributor of addresses on the 2000 MAF (Vitrano et al., 2003), so too is it reasonable to expect that the 2000 MAF will contribute the core set of addresses to the 2010 MAF.

Still, having the 2000 MAF in hand does not give license to defer active address list updating to 2009. We believe—and sincerely hope—that the sentence is a misstatement; indeed, in later text, the baseline design document strikes a more reasonable note, pledging "work with USPS, local, and tribal partners" through the decade and saying that, "whenever we identify new housing units or those that no longer exist, we will update our files" (Angueira, 2003b:4). A different interpretation of the first sentence is that the general term "address list updating" is being used to describe a more specific operation, most likely block canvassing.

The draft baseline design continues (Angueira, 2003b:3–4):

> As part of the MAF/TIGER Enhancements Project, the Local Update of Census Address program (LUCA) will have been streamlined and improved based on lessons learned from the Census 2000 LUCA experience, and the address list for the entire universe will have been maintained and updated on a continuing basis. . . . There will be an address updating operation in 2009 in areas that we believe have experienced significant changes. . . . The streamlined, ongoing LUCA program will culminate with a final opportunity for local governments to review their address lists, which will occur prior to address canvassing. We will then validate any LUCA adds during address canvassing. We will have a New Construction operation, and will attempt to include those addresses in questionnaire delivery. The New Construction adds will be validated during a later operation.

The implications of these statements are disturbing in two key respects. First, the passage lists several different address updating mechanisms (considering updates from the DSF and CAUS as part of the MAF being "maintained and updated on a continuing basis") but provides a very weak sense of their order and scheduling. That a block canvass would not overlap a LUCA-type program—as it did in 2000—is an improvement. But how all the activities fit into a coherent timeline is not clear—particularly if 2009 is the start date. Second, the casting of the block canvassing operation as a validation step for LUCA is troubling as it imparts to block canvassing a "most trusted" authority. We do not argue that local and tribal knowledge of addresses is foolproof, and there is need for some sort of validation. However, it is unclear whether empirical evidence supports the assertion that block canvassing is more likely than other operations to correct addresses. Creating the impression that, near the end of the decade, the Census Bureau will make a major deployment of field staff to perform block canvassing because local input on address information is somehow less trustworthy may only serve to further hinder participation by local and tribal authorities in Census Bureau activities.

3–E RECOMMENDATIONS

3–E.1 Plan MAF Improvements Independent of MAF/TIGER Enhancements

The Census Bureau needs to outline goals pertaining directly to the MAF independent of the goals for TIGER—for example, in the development of quality metrics and the identification of housing unit duplication. Overall milestones and tasks need to be specifically set for Objectives Three and Four, to determine how these objectives may work to control housing unit duplication and to more accurately identify and account for multiunit housing structures. It is also vitally important that MAF improvements be coordinated with efforts to list and enumerate the population living in special places and group quarters; we will describe both group quarters and multiunit structures in Chapter 5.

The Panel to Review the 2000 Census discusses the problems of the 2000 MAF in great detail (National Research Council, 2004:Ch.4), and argues that the process for updating the MAF during the years leading to the 2010 census is in need of serious revision. We concur, and accordingly stress the following recommendation (a synthesis and extension of both Recommendation MAF–1 from our second interim report and National Research Council (2004:Rec. 4.1)):

> ***Recommendation 3.1:*** **The Census Bureau must devise a plan and develop effective procedures for updating and correcting the Master Address File (MAF). A complete and accurate Master Address File is critical not only to the success of the 2010 census but also to the effective implementation of the American Community Survey, the other household surveys conducted by the Census Bureau, and the 2008 dress rehearsal. Because the 2000 MAF was not simply discarded following the 2000 census (as occurred in censuses prior to 1990), the 2010 census will have as a base an address file of unprecedented completeness, but that does not obviate the need for continual updating, filtering, unduplicating, and cleaning of the MAF during the years leading to the 2010 census.**

The plan for a continually updated 2010 MAF must include, but not be limited to, the following:

1. **A clear articulation of how the MAF/TIGER Enhancements Program and other Census Bureau activities will add missing housing unit addresses, remove duplicate addresses, and generally correct the Master Address File, independent of benefits derived from being cross-referenced to an updated TIGER database;**
2. **More effective definitions of housing units and methods to obtain accurate address listings for structures containing multiple housing units, as it is not sufficient to know only the address or geographic coordinates of the structure location;**
3. **Detail on the temporal sequencing and adequacy of address updates from the U.S. Postal Service's Delivery Sequence File, the Census Bureau's Community Address Updating System, and as-yet unspecified local partnership programs;**
4. **More effective means to define, list, and enumerate group quarters living arrangements, which should be done in coordination with the development and maintenance of the MAF; and**
5. **A detailed plan for Objective Five (quality metrics) of the MAF/TIGER Enhancements Program, including a program of evaluation and assessment of MAF coverage and input to the MAF/TIGER redesign (Objective Two), so that the revised database structure includes appropriate address source codes and other useful variables for evaluation.**

3–E.2 Coordinate Responsibility for the MAF

In Chapter 6, we advocate the creation of a new position within the Census Bureau—a system architect for the decennial census—with the primary goal of integrating and coordinating work on architecture remodeling. We believe that improving the MAF is likewise an area that would benefit greatly from focused staff effort. At least three major divisions within the Bureau (Geography, Field, and Decennial Management; see Box 2.2) have a strong stake in the maintenance and use of the MAF as it pertains to the decennial census, and the Demographic Surveys division also has a stake given MAF use in conducting the Bureau's household surveys. Given the legitimate (but sometimes competing) interests of the various divisions, it would be useful to vest responsibility for coordinating MAF improvement and research in one office with both the connections and the ability to work with all relevant divisions.

We reiterate a recommendation from our second interim report (National Research Council, 2003a:Rec. MAF–2):

> ***Recommendation 3.2:* The Census Bureau should create and staff a position to oversee the development and maintenance of the MAF as a housing unit inventory, with a focus on improving methods to designate, list, and update units. This position should be responsible for development and implementation of plans drawn up consistent with Recommendation 3.1.**

Census Bureau staff expressed skepticism about this recommendation in their reaction to the second interim report at the panel's final public meeting in September 2003, arguing that the Bureau's organization is not given to the creation of centralized "czar" positions. That argument, however, underscores the point of this and several other recommendations in this report: real integration in achieving census objectives will require some thinking outside the lines of existing organizational trees. In our assessment, the Census Bureau's approach of handling MAF issues by committee is ineffective and leads to serious underutilization of the Bureau's existing staff and resources; MAF development should be supported with a clear structure of organization and accountability, as the Enhancements Program has done for TIGER.

3–E.3 Improve Research on the Delivery Sequence File

Our next four recommendations call for the development of an empirical, research-based approach to MAF updating efforts throughout the decade. Each prospective address input source should be carefully examined, weighing strengths, weaknesses, and costs, and reasonable estimates of the source's potential contribution to the 2010 MAF should be produced.

The U.S. Postal Service's Delivery Sequence File provides twice-yearly "refreshes" to the MAF under the Census Bureau's current geographic support system. The efficacy of these updates—in general, and differentially by geography and urban/rural status—has not yet been fully demonstrated.

> ***Recommendation 3.3:*** **The Census Bureau should pursue more effective partnership and research collaboration with the U.S. Postal Service, including but not limited to further work on "undeliverable as addressed" items from the 2000 census, assessment of the address coverage quality of the Delivery Sequence File (DSF), and possibilities for more accurate translation of post office box listings and other DSF entries to street addresses and geographic coordinates.**

3–E.4 Define the Role of the Community Address Updating System

Objective Four of the MAF/TIGER Enhancements Program—CAUS—has been delayed in implementation due to the lack of initial funding for the ACS. The expectations for CAUS have never been entirely clear. As we noted in Section 3–D.1, the system has been described as vital to securing address updates "in rural areas with non-city-style addresses," which represent approximately 15 percent of the population (U.S. Census Bureau, 2003c:9). However, at the panel's last public meeting in September 2003, senior Census Bureau staff commented that not much would be lost if CAUS were not fully implemented, contingent as it is on funding of the ACS.

However, the Census Bureau has indicated that it trained 400 field representatives on CAUS methodology in August 2003, began listing operations in October 2003, and has continued to refine the ALMI GPS-equipped laptop computer used to collect CAUS data (U.S. Census Bureau, 2003e:4). In light of these investments, as 2010 planning proceeds, the Bureau needs to make clear the expectations for CAUS, including assessment of the long-term feasibility of the activity and of its potential contribution to the 2010 MAF. If CAUS is indeed crucial to securing updates from rural areas, given the uncertainty about the program's implementation, consideration needs to be made as to whether alternate sources could provide the information.

> ***Recommendation 3.4:*** **The Census Bureau should assess how critical the Community Address Updating System (CAUS) is to providing address updates in rural, non-city-style address areas. Such an assessment should include not only estimates of the number of addresses that could be provided and the workload that could be handled by CAUS/American Community Survey staff, but also empirical evidence on coverage gaps in the U.S. Postal Service's Delivery Sequence File by geographic area or type.**

3–E.5 Plan Local Geographic Partnerships and Implement Early

To its credit, the Census Bureau has recognized the importance of partnerships with local and tribal governments by designating their creation and maintenance as Objective Three of the Enhancements Program. The Bureau's RFP for the TIGER realignment of Objective One makes this clear, noting that "the success of [Objective One], and the continuous update of the information in MAF/TIGER, requires ongoing interaction between the Census Bureau and its federal, state, local, and tribal government geographic partners." However, the Bureau has not provided a clear indication of how such partnerships would work.

While the panel acknowledges that the funds available for expanding and encouraging geographic partnership options have been limited, the cryptic descriptions of Objective Three that we have received do not make clear how and when the Bureau intends to involve local and tribal partners in these programs.

A major stated role for local and tribal geographic partners is to contribute to Objective One by sharing their current GIS files with the Census Bureau to support TIGER realignment. But in this matter, and in past geographic interactions such as LUCA, the Census Bureau has often approached "partnership" as a one-sided exchange: "partners" expend resources and turn information over to the Bureau. The principal reward to a local or tribal government for entering into such a partnership is definitely not trivial: the prospect of a more accurate census count. The panel recognizes that the Census Bureau is not a fund-granting organization and hence cannot directly subsidize local or tribal governments to improve and submit their geographical resources. That said, the Bureau should aim for partnerships that are true exchanges of information: for instance, by giving census field and regional staff an increased role in interacting with local and tribal authorities and collecting information updates. At the very least, steps should be taken to lessen the burden of partnership on the local and tribal governments—for example, by conducting LUCA-like address list reviews electronically with submissions via the Internet, and coordinating the various geographic data collection programs so that localities are not asked for similar information in different formats by different divisions of the Census Bureau.

The Census Bureau needs to articulate a plan for communication with localities that takes advantage of existing structures, including the State Data Center Network, the Federal-State Cooperative Program for Population Estimates, state and regional councils of governments, and other local governmental entities. The role of the Census Regional Office Geographic Coordinators relative to these entities and to Census Bureau headquarters needs to be spelled out.

The ability and willingness of different governments to join forces with the Census Bureau vary widely. It is inevitable that local efforts will be differentially expressed in different areas of the country, whether such efforts involve mapping, address listing, or the nurturing of partnerships. While all areas should receive equal treatment in the spirit of fairness, local interest, feasibility, and cost-effectiveness might well dictate otherwise. Moreover, although geographic partnerships with local and tribal governments can be useful to tap the knowledge and expertise of those closest to the field, variations in GIS usage may affect the accuracy of local and tribal government geographic resources and may introduce errors when combined with census resources. In the interest of effectiveness, we recommend careful analysis of the successes and failures of prior LUCA programs in order to properly conduct future community participation programs. Close evaluation of the 2000 address file by type of enumeration area, by dwelling type, by the contribution of geographic update programs like LUCA, and by region of the country—highlighting areas where elicitation of local and tribal information may be most beneficial—is surely required if the Census Bureau is going to maintain the MAF in a cost-effective manner in the years leading to the 2010 census. The Bureau's future plans for LUCA and other partnership programs should also provide for evaluation of those partnerships, not only to inform the effectiveness of local contributions from the census perspective but also to give feedback to participating local and tribal governments.

We reiterate a recommendation from our second interim report (National Research Council, 2003a:Rec. MAF–3) and add two other points on the nature of partnerships:

> ***Recommendation 3.5:*** **The Census Bureau should immediately develop and describe plans for partnerships with state, local, and tribal governments in collecting address list and geographic information. Such plans should include a focus on adding incentives for localities to contribute data to the census effort, making it easier for localities and the Bureau to exchange geographic information. Accordingly, plans for partnerships should include:**
>
> - **clear articulation of realistic schedules for local input and review;**

- **definition and clear presentation of benchmark standards for local data to be submitted to the Bureau;**
- **mechanisms for providing effective feedback to local and tribal governments, detailing and justifying the Bureau's decisions to use or not use the information provided; and**
- **coordination of efforts across the Bureau so that calls for local and tribal entities to supply input to the Master Address File, TIGER, the Boundary and Annexation Survey, and other Bureau programs are not unduly redundant and burdensome.**

3–E.6 Justify the Complete Block Canvass

In Section 3–D.2, we commented on Census Bureau reaction to the assumption, stated in our second interim report, that the Bureau hoped to forestall a complete block canvass in the 2010 census. Our commentary in the interim report continued (National Research Council, 2003a:66):

> In the absence of evidence that the combination of DSF and LUCA leading up to 2010 can overcome the last-minute doubts that arose in the late 1990s and without a clearer plan for CAUS—it is difficult to see how a full block canvass can be averted.

We continue to stand by this assertion, and have called for development of empirical evidence on possible DSF, CAUS, and LUCA contributions to the 2010 MAF. Likewise, we believe that the Census Bureau's decision to proceed with a full block canvass should also be justified with empirical evidence.

We do not suggest that block canvassing is an idea that lacks merit. The evaluations of the 2000 census suggest that the effort contributed many addresses to the MAF (Burcham, 2002; Vitrano et al., 2003) and was generally very good at verifying existing units. However, evidence also suggested relatively high rates of inconsistency (22–24 percent) between addresses added or deleted by the block canvassing operation and results in the census; an example of an inconsistency is a housing unit added by block canvassing but then found during the census operations to be an invalid housing unit (Burcham, 2002:38–39).

We believe it rash to commit to such an expensive operation as full block canvassing absent both a compelling base in empirical evidence and a determination that targeted canvasses in specific (e.g., fast-growth suburban) areas are infeasible. It is decidedly a mistake to consider a full block canvass without early attention to effective canvass techniques for all types of housing stock, particularly small multiunit structures (see Section 5–C.1). The panel is also concerned that reliance on a block canvass may send unfortunate mixed messages about the relative quality of the address list needed for different purposes—that special operations are needed to derive an address list of presumably higher quality than that needed for the Census Bureau's other surveys and, particularly, the ACS. We therefore recommend:

> ***Recommendation 3.6:*** **The Census Bureau should evaluate the necessity of its plans to conduct a complete block canvass shortly before the 2010 census. Such justification must include analysis of extant census operational data and should include, but not be limited to, the following:**
>
> 1. **arguments as to why selective targeting of areas for block canvass is either infeasible or inadequate, and as to how the costs of the complete block canvass square with the benefits; and**
> 2. **analysis of how a full block canvass fits into the Census Bureau's cost assumptions for the 2010 census.**
>
> **If plans proceed for a complete canvass, the Bureau should also consider how such a mass field deployment prior to 2010 could be used to achieve other improvements or efficiencies, such as the collection of GPS trace data as supplement to or as quality control for the TIGER realignment.**

3–E.7 Exploit 2000 MAF Data, and Redesign MAF for Evaluation in 2010

A recurrent theme in our preceding remarks is that there is a strong need for empirical assessment of the quality of potential address sources for the 2010 MAF. The natural starting place for such an evaluation would be the Census Bureau's MAF Extract. Based on the 2000 census Decennial Master Address File—the "snapshot" of the MAF used to generate census mailing labels and to monitor mail response—the MAF Extract includes "flags" that indicate which of several sources contributed the address to the MAF. The MAF Extract also contains selected outcome measures, such as whether the address record was actually used in the 2000 census and whether it was tagged as a potential duplicate during the ad hoc duplicate screening program of early to mid-2000 (Nash, 2000).

The MAF Extract has certain liabilities, chief among them that the system of flags used to indicate the source of an address does not show the complete history of an address in the MAF. Other than a rough temporal ordering of the input sources themselves, it is usually impossible to determine which source first contributed the address. Nonetheless, the extract is critical to answering key questions about the MAF-building process, and the panel continues to urge that the data resource be tapped for as much information as possible.

Analyses of the MAF Extract should consider the type of enumeration area for each address in the 2000 census (e.g., mailout/mailback or update/leave) as well as geographic region. The main objective of the analysis is not to highlight how different areas of the country may have fared under various programs, but rather to obtain knowledge of how people in those areas respond and interact with census activities in order to improve planning for future census programs.

Some key questions to address through Census 2000 evaluations are the following:

1. Why were addresses included in the MAF but not in the 2000 census?

   This question provides perspective for the others on this list and is a good starting place.

2. How useful were the DSF updates in the identification of new units, especially in high-growth areas of the nation?

   The goal is to examine how much of the newest housing was picked up in a timely fashion by the U.S. Postal Service. The answers can provide valuable clues about the effort the Census Bureau should put into other avenues (e.g., new construction programs) as sources of information on new housing.

3. How effective were LUCA inputs relative to what was already known (or was promptly seen) in a DSF update? Of those contributions that can be determined as "unique," how many governments were represented and what kind of housing do these addresses represent?

   While LUCA must be conducted as part of the preparation for the 2010 census, the resources the Census Bureau chooses to expend on it can vary dramatically. The answer to this question can also inform strategies for the LUCA program for 2010.

4. What were the original sources of address records that were deleted in the ad hoc duplicate identification and removal process conducted in 2000?

   Duplication related to address listing anomalies can be rectified once the specific problems with the duplicate addresses have been identified. Identifying the original source of the affected addresses is a prime means for doing that.

5. What were the original sources of addresses that were flagged as potential duplicates but later reinstated?

   This question addresses the hypothesis that some addresses, originally considered as potential duplicates, were put back into the census in error. The Census Bureau already has an estimate of this number. By identifying the original sources of these addresses, the Bureau will have valuable clues about what produced this problem and how to avoid it in the future.

6. What were the original sources of addresses for housing units where an interview was not obtained in nonresponse follow-up (NRFU)?

One hypothesis about the shortfall of long-form data in the 2000 census posits that NRFU enumerators encountered high levels of resistance from respondents who were being enumerated for the first time (some of whom escaped detection in 1990). Where did the addresses of these tough-to-enumerate units fall? (Of course, this is not the only or most likely hypothesis to explain problematic long-form data, but the question warrants attention and the Census Bureau's MAF Extract data may be able to provide useful information.)

7. What were the original sources of addresses for housing units that were subsequently declared nonexistent or were not found in NRFU?

   NRFU enumerators had the option of entering codes for "cannot locate," "duplicate," and "nonresidential," among others, as reasons for listing a unit as "nonexistent." Were these potential duplicates added back in, were erroneous addresses brought in from LUCA that were not detected by the Census Bureau, or were these problem addresses disproportionately from some other original source?

8. For cases where a unit was determined not to exist in coverage improvement follow-up (CIFU; the final follow-up stage during the actual fielding of the census), what was the original source of the address? How many addresses were erroneously kept in the census and then deleted when the Bureau went out to check in CIFU?

The Census Bureau's topic report on address list development (Vitrano et al., 2003) is a step toward answering these questions. In particular, it makes strides toward managing the poor and confusing MAF codes indicating operations that added or edited the address in order to ascertain the original source of each address record. But it is only a step. Accordingly, we recommend:

***Recommendation 3.7:* The Census Bureau must:**

- **fully exploit the address source information in the MAF Extract in order to complete 2000 cen-**

sus evaluations, fill gaps in knowledge remaining from the 2000 census evaluations, and assess causes of duplicate and omitted housing units; and

- **build the capability for timely and accurate address evaluation into the revised MAF/TIGER data architecture, including better ways to code address source histories and to format data sets for independent evaluation.**

CHAPTER 4

American Community Survey

THE CONSTITUTIONAL MANDATE for the decennial census is to provide a basic head count for purposes of apportionment, but the nation's need for accurate measures of its civic health has led the census to develop well beyond a simple count tabulated by age, race, and sex. Over time, the roster of questions included in the census expanded to cover a wide array of socioeconomic and demographic characteristics. The emergence of the statistical theory of survey sampling in the early 20th century brought with it the potential to collect detailed characteristics information without unduly burdening the entire American public. Asking detailed characteristics information from only a sample of the populace began in the 1940 census, when six questions on socioeconomic status were asked of only 5 percent of respondents. In 1960, the concept took its next evolutionary step when two separate census forms began to be used, a design feature that continued through the 2000 census. The short form covers the basic information items to be asked of all residents; the long form—administered only to a sample of the public—includes the complete battery of characteristics questions. In 2000, for example, the short form contained queries for six basic items—age, sex, Hispanic origin, race, relationship to census respondent, and housing tenure (renter/owner), while the long form administered to roughly one-sixth of households added about 62 items, 36 of them pertinent to demographic and economic characteristics and 26 related to housing.

Fifty years after the development of separate short and long forms in the census, the Census Bureau proposes to make another change in the collection of population characteristics data by introducing the American Community Survey (ACS). A major household survey intended to include 250,000 housing units each month, the ACS would replace the decennial census long-form sample and permit continuous measurement of the same data items currently collected only every 10 years on the census long form. The 2010 census would therefore include only the short form, which would enable easier (and potentially more accurate) data collection in the census and save costs on data capture from completed paper questionnaires. At the same time, the data on characteristics currently collected on the census long form would be produced on a more timely basis, offering annual assessments rather than a static once-a-decade snapshot.

The potential rewards of the ACS are great, but so too are its inherent risks. The survey's success is contingent on sustained long-term funding, and year-to-year fluctuations in allocated spending levels could cause severe data quality problems, particularly for small population groups. Estimation based on continuous measurement such as the ACS—most likely making use of moving averages of several years of data—also raises conceptual and feasibility issues that must be addressed in order for the survey to win support. These risks, and others, are significant, but perhaps the most important risk associated with the ACS is simply one of timing. A final decision on the methodology for the 2000 census was reached dangerously close to Census Day; extended delay in reaching agreement at all levels—the Census Bureau, the administration, and Congress—about the role of the ACS could similarly raise the risk of having to revamp census design very late in the cycle. The decision on whether the ACS will proceed in full—and, with it, determination of the fate of the census long form—is the single most important element in terms of defining the general shape, structure, and design of the 2010 census.

In this chapter, we discuss the background of the ACS and describe the current plans for a fully operating ACS in Section 4–A. We then begin our assessment by identifying key questions (4–B); these major questions generally center around the challenges of estimation using the ACS (4–C) and the basic quality of ACS data (4–D). Our general summary and assessment of the ACS' proposed role in the 2010 census (4–E) is followed by an outline of major features required in the intensive research and evaluation effort that should complement ACS operations (4–F).

4–A BACKGROUND AND CURRENT PLANS

Work on what is now known as the American Community Survey commenced after Alexander (1993) revisited the idea of a continuous measurement survey for gathering long-form data as a complement to a short-form-only census (for historical context, see also Hauser, 1942; Kish, 1981, 1990). Two previous National Research Council panels supported the general principle of a continuous measurement survey and urged further research (National Research Council, 1994, 1995); however, National Research Council (1995) concluded that a proposal to implement the survey to replace the census long form in the 2000 census was infeasible, given inadequate lead time and unresolved conceptual problems. The ACS was also the focus of a 1998 National Research Council workshop to discuss research priorities (National Research Council, 2001b).

4–A.1 Test Sites and the Census 2000 Supplementary Survey

Though the ACS was ruled out as a replacement for the long form in 2000, the mid-1990s burst of research and writing about the prospects of continuous measurement launched a wider research and evaluation effort. Pilot data collection for the ACS began in four test sites in 1996. By 1999, data collection in this demonstration phase had grown to include thirty-one sites across thirty-six counties (U.S. Census Bureau, 2003c). During the initial pilot phase in 1996–1998, residents were sampled at a markedly higher rate—15 percent, increased to 30 percent in some communities—than is planned for the full-scale ACS.

More significantly, as the ACS began to be adopted as part of the developing 2010 census plan, an experiment was developed in conjunction with the 2000 census to attempt to address the basic question of operational feasibility (that is, whether it is possible for the Census Bureau to conduct the decennial census and an ongoing survey containing usual long-form items at the same time, both operationally and in terms of burden on respondents). Accordingly, the Census 2000 Supplementary Survey (C2SS) began in January 2000 and continued data collection through December 2000. This prototype ACS sampled from 1,203 counties and covered approximately 700,000 households over the course of the year. Data collection continued at these levels in 2001–2003. A report prepared as part of the 2000 census evaluation program concluded that operating a large continuous measurement survey in parallel with the decennial census was operationally feasible, based on the 2000 census and C2SS experience (Griffin and Obenski, 2001).

Original plans called for the ACS to begin full field implementation in 2003, a schedule that would support publication of small-area estimates in 2008. However, congressional stalemate on the budget for fiscal year 2003 delayed full implementation by at least one year.

4–A.2 Current ACS Implementation Plans

Under the funding levels appropriated for fiscal 2004, questionnaire mailing for a full-scale ACS would begin during the fourth quarter (July–September) of fiscal 2004. Follow-up field work would be deferred until after September 2004, pushing the considerable expense of field interviewing into the fiscal 2005 budget process. Prior to the fourth quarter mailing, data would continue to be collected in the thirty-one test sites and at the C2SS levels (Lowenthal, 2003a).

When the ACS is fully fielded, it will use as its sampling frame the same Master Address File (MAF) used by the decennial census. The annual sample of housing units chosen for participation in the survey will be divided into monthly mailout panels, each of which will be a systematic sample across the complete address list. Thus, it is intended that each month's sample will be a representative sample (approximately $\frac{1}{480}$) of the population of each area of the United States. In practice, this simplified sample selection process will be modified by practices similar to those used for the decennial census long form, including oversampling of small geographic areas.

The ACS is intended to be administered primarily via mailout/mailback. However, the proposed ACS techniques to follow up with households that do not return the mail form differ from decennial census practice. All mail nonrespondents will be initially followed up by computer-assisted telephone interviewing (CATI) during the month following questionnaire mailout, if there is an available phone number. After CATI follow-up, a random one-third of the remaining nonrespondents will be designated for follow-up by field enumerators using computer-assisted personal interviewing (CAPI). The precise nature of this sequential follow-up process remains to be determined; there are tentative plans to sample areas with low mail and telephone response rates at a higher fraction rather than a strict one-third random sample.[1] This oversampling may help to make sample variances more comparable across areas.

The stagewise nature of ACS follow-up leads to another important design feature, which is that all of the information *collected* in a given month will be used as inputs for that month's estimates. That is, a particular month's estimates may include mailback responses from the present month's systematic sample of housing units as well as completed telephone and personal interviews from one and two months prior, respectively. This design choice is advantageous in that it simplifies data processing and production load—there is no need to wait until month $t+2$ for final resolution of all the housing units chosen in month t before processing responses already submitted. But it does raise complex methodological challenges, including the choice of weighting methods to address unit nonresponse.

While the size of this survey will make possible some direct small-area estimates, the estimates for areas with a population

[1] Due to budget constraints, the Bureau may be required to reduce the sampling rate in *higher* mail and telephone response areas to accommodate this oversampling. The implications of such a shift need to be researched ahead of time before plans are finalized.

of less than 65,000 typically will be produced by aggregating information over either 3 or 5 years, depending on the size of the area. At this time, moving averages are planned to be used for these aggregate-year estimates, though other possibilities could be considered in the future.

The need for a 5-year window to produce detailed small-area estimates puts a firm constraint on the date of full ACS deployment. The initial plans for full deployment in 2003 would have produced small-area estimates in 2008, allowing some time for the new ACS figures to gain acceptance as a long-form replacement. To match the long-form data production schedule of the 2000 census, the absolute deadline for full (and sustained) implementation of the ACS is 2007, which would permit the publishing in 2012 of national estimates analogous to those from the long form.

4–B ASSESSING THE ACS

In simplest terms, the root practical question that must be answered in justifying the ACS is whether the information generated by the survey is an adequate replacement for the data currently collected on the census long form. Parsed at the most basic and literal level—whether the ACS and the long form are substitutable in content—the answer is simple. By design, the ACS covers the same topics and data items as the census long form; exact question wording and ordering may vary, but in general terms the content matches. Thus, in the simple sense of topical content, the ACS is an obvious substitute for the long form.

The more challenging question is whether the ACS can replace the census long-form sample in terms of performance and function. This basic question can be further subdivided into key subquestions, the answers to which are vital to bolstering the case for the ACS.

- For all but the largest population or geographic groups, ACS estimates will be based on averages across multiple years of data. Is the ACS able to satisfy all of the needs currently addressed by long-form data, or are there applications based on the census long form for which substitution of a moving average-type estimate from the ACS would be inappropriate?

- How well will ACS estimates match other estimates of the same phenomena? That is, how will ACS measures compare in level or trend to traditional long-form estimates or to other survey measures?
- What is the quality of ACS estimates and data relative to the census long form? Specifically, what can be said about error—both bias and variance—and undercoverage in data collected through the ACS, and how do they compare with those incurred through the census long form?

These and other questions involve concerns about methods of estimation based on the ACS and about the inherent quality of ACS data and estimates; we offer more detailed comment on these concerns in the following two sections.

4–C ESTIMATION USING THE ACS

4–C.1 Adequacy of Moving Averages as Point Estimates

A basic concern about the American Community Survey as a replacement for the census long form is whether ACS estimates—which, particularly for small areas or groups, would be moving averages of multiple years' data points—can effectively replace fixed-point-in-time estimates. Specifically, the concern is whether fund allocation formulas or other public and private planning needs for demographic data can be addressed using a combination of data from multiple years. The Census Bureau has issued a draft report that attempts to address users' concerns about this shift (Alexander, 2002), and Zaslavsky and Schirm (1998, 2002) outline the advantages and disadvantages a locality may experience through use of either a moving average or a direct (census) estimate.

The crux of the debate on this point is that a moving average is a smoothed estimate; by averaging a particular data observation with other observations within a particular time window, the resulting estimate is meant to follow the general trend of the series but not be as extreme as any of the individual points. The ramifications of this method emerge when moving average estimates are used in sensitive allocation formulas or compared against strict eligibility cutoffs. A smoothed estimate may mask or smooth over an individual-year drop in level of need, thus keeping the locality eligible for benefits; conversely, it may also mask individual-year spikes in need and thus disqualify an area from benefits. It is clear that the use of smoothed estimates is neither uniformly advantageous nor disadvantageous to a locality; what is not clear is how often major discrepancies may occur in practice.

One answer to this conundrum is to use sample-based estimates from individual years instead of moving averages. These estimates would be unbiased in terms of probability but could be highly variable, which would affect aspects of formula grants such as "hold-harmless" provisions.[2] A related worry that has been expressed about moving averages is that, by incorporating estimates from other time periods, the estimates for a given period could be substantially biased and not truly reflect the conditions for that period. The empirical challenge is to assess the bias that may result from averaging over 3 years of data compared to 5, and try to weigh the magnitude of that bias against the bias associated with using an up-to-12-years-old long-form estimate. Intuitively, it is sensible that, when examining data series in which change is substantial between decennial census years, moving average estimates would be preferable to seriously outdated estimates. When there is little change through the decade, there should be little difference between the two estimates. However, since this is an empirical question, the Census Bureau should carry out research that helps to evaluate this trade-off.

The continuous measurement properties of the ACS give it unique advantages over the decennial snapshots available from the census long form, but they also raise another, related point of

[2]A "hold-harmless" provision in a funding formula is one that limits the amount by which an allocation can change from one year to another; for instance, under a 70 percent hold-harmless level, a unit's allocation may only decrease by up to 30 percent. In a hold-harmless situation, an unusually volatile observation one year due to increased variability could mean that the unit's allocation may remain out of true alignment for several cycles due to the amount of allocation automatically carried over.

concern regarding moving averages: assessment of year-to-year change in a data series. It is incorrect to use annual estimates based on moving averages over several years when assessing change since some of the data are from overlapping time periods and thus identical. At the least, the results will yield incorrect estimates of the variance of the estimates of change. Therefore, users should be cautioned about this aspect of the use of moving averages. Along the same lines, moving averages present the same types of problems when they are used as dependent variables in various statistical models, in particular time-series models, and in some regression models. Therefore, the Census Bureau could bolster the case for the ACS and potentially help relieve users' concerns if it produced a user's guide that details the statistical uses for which moving averages are and are not intended, the problems they pose to users, and the means to overcome them.

4–C.2 Comparing ACS/C2SS to the Census Long Form

Thus far, we have outlined from conceptual and theoretical perspectives the issues surrounding the adequacy of ACS estimates to replace the long form. It is also natural to address the question from a more pragmatic point of view: the ACS and the census long form purport to measure the same basic phenomena, but do the resulting data from both series actually tell the same story?

Comparisons of how the ACS or C2SS estimates match census long-form estimates implicitly treat the census long-form data as an effective "gold standard"—a questionable assumption at best, given that it discounts the various (and sometimes substantial) sources of error to which the long form is subject. First, the long-form data for small areas are subject to substantial sampling error. In addition, as mentioned above, the long form is particularly subject to nonresponse, and for some sample items the amount of nonresponse for the long form in the 2000 census was extremely high (National Research Council, 2001a, 2004).

Love (2002) has identified a number of sources of differences between the ACS (or C2SS) and long-form census estimates that complicate any direct comparison. These include differences in: reference dates; modes of nonresponse follow-up; criteria used to decide if a response is acceptable; edit and imputation techniques; methods for data capture and processing; the use of proxy interviews (they are accepted for the decennial census but not by ACS); definition of respondent eligibility; and weighting procedures used to address nonresponse and sampling (e.g., the weighting of the long-form estimates to the basic complete-count data). The reference period associated with a question item is of particular interest for ACS estimates, since annual averages will be the average of responses corresponding to twelve different reference periods, depending on when the questionnaire was applied. There are also differences in the target population; for example, the ACS does not currently include group quarters in its survey, but the census does.

Work on comparing the ACS (test sites) and C2SS estimates to census long-form estimates has been initiated by the Census Bureau. To date, what is known is that there are some substantial differences. Generally, these differences can be explained by the amount of sampling error in the two surveys (U.S. General Accounting Office, 2002a); however, examination of C2SS data suggests significant differences for the number of housing units lacking complete plumbing facilities and for the number of unpaid workers in a family, for instance. At the state level, a large number of C2SS estimates differed from the long-form estimates by at least 10 percent, including the number of workers that commute using public transportation, the number of households with income above $200,000, the number of housing units that lack complete plumbing facilities, and the number of renter-occupied units with gross monthly rent of $1,000 to $1,499.

The Census Bureau needs to complete this analysis, including the contribution of sampling variance, for all years of data collection, and attempt to identify the sources of differences other than sampling error. A priority of this analysis should be responses related to residency, but all responses should be examined.

4–D QUALITY OF ACS ESTIMATES

The error associated with ACS data may be decomposed into sampling error (sample variance) and nonsampling error, the latter of which can be further separated into error due to nonresponse and measurement error due to various causes.

At the most basic level, sampling error in the ACS will be slightly larger than that for the long-form sample because the total ACS sample size over a 5-year period will be slightly smaller than that for the census long form. On its own, this difference is unlikely to have a substantial impact on users. However, sampling error due to initial mail and CATI nonresponse is widely variable and could be appreciable in some small areas.[3] As a result, the Census Bureau is considering raising the sampling rate for CAPI follow-up for areas with high mail and telephone nonresponse to make this source of sampling error more comparable across areas.

It should be noted as we review these issues that, generally, these concerns are generic to all surveys, including the census long form—that is, the concerns are not raised as specific flaws of the ACS. They are, nonetheless, features of the ACS that must be measured and weighed in deciding how best to use the data.

4–D.1 Estimating Nonresponse

Unit Nonresponse

One part of nonresponse in a survey program like the ACS is unit nonresponse—that is, failure to obtain questionnaires and data from households selected for inclusion in the sample. A common combined measure of unit nonresponse and survey undercoverage is the sample completeness ratio, which is the sample-weighted estimate of the population count for a certain area divided by the census count for the area. The sample completeness ratio nationally for C2SS was 90.2 percent, while the comparable figure for the 1990 long-form sample was 89.7 percent (U.S. Census Bureau, 2002b). These figures may appear close, but some care must be taken in interpreting them. For example, the long form accepts proxy responses from landlords or neighbors while proxies are not permitted in the ACS or C2SS, and it is generally accepted that proxy responses are of lower

[3]See Salvo and Lobo (2002) for relevant discussion on this point.

quality than responses by household members.[4] So the programs, and these ratios, are not directly comparable. Still, the C2SS seems to be roughly equivalent to the long form with respect to unit nonresponse and survey undercoverage.

Another statistic that is often examined to assess the quality of survey data collection is the rate of mail questionnaire return. This is because, in the census context, information collected through self-response is typically considered to be of higher quality than information collected through field enumeration (National Research Council, 1995). For the C2SS, the mail return rate was 51.9 percent (U.S. Census Bureau, 2002b:11), lower than the 71 percent for the 2000 census long form (National Research Council, 2004:100). For the 2000 ACS in the Bronx County test site, the mail return rate was 36.4, compared to 55.8 for the long form in the 1990 census (Salvo and Lobo, 2002). While this difference could contribute to a lowering of the quality of ACS information relative to census long-form information, it might be addressed by improved field data collection.[5]

Salvo et al. (2003) apply a metric of minimal completeness to specially prepared operational data from the 2000 ACS and the 2000 census for Bronx County, New York (one of the thirty-one test sites.[6] They found that 49 percent of enumerator returns

---

[4]Nonresponse follow-up for census long-form data was often concluded with the collection of short-form data only (that is, a premium was placed on gathering the basic short-form characteristics from as many nonrespondents as possible rather than insisting on a complete long-form return). Such forms are treated as long-form unit nonresponse.

[5]As Salvo and Lobo (2002) demonstrate, there is substantial heterogeneity in the ACS mail return rate and in other measures of nonresponse as a function of characteristics often associated with being difficult to count in the census. Therefore, it should be understood that both the ACS and the census long form are more or less successful in collecting quality data depending on the area of interest.

[6]Salvo et al. (2003) use a measure of completeness similar to that used in census processing to determine if a household data record is complete enough to be considered "data-defined." To be complete under this metric, at least one member of the household had to have answered two of the basic complete-count items asked on the census short form (e.g., age or sex) and two of the sample data items asked on the census long form. Preliminary results of the Salvo et al. (2003) analysis were reported in Salvo and Lobo (2002).

for the long form failed the completeness test, whereas only 14 percent failed for the ACS.

The overall weighted survey response rate for the ACS has been calculated as 95.4 percent, which is very high for a household survey. This rate includes responses across the different possible modes of administration (telephone, in-person) but does not factor in survey undercoverage.

Item Nonresponse and Invalid Response

Extant research based on item imputation rates for responding households not only measures item nonresponse but also includes imputations for responses that fail consistency edits. However, this complication is relatively infrequent and is consistently applied to both the C2SS and the census long form. As a result, we feel that it is reasonable to compare item imputation rates to measure the impact on data quality from item nonresponse. Item imputation rates for the C2SS were substantially lower than those for the 2000 census for the basic data items on both the short and long forms. For example, for age, the census imputation rate was 3.6 percent, whereas for the C2SS it was 2.4 percent. Salvo and Lobo (2002) report that the allocation rate (essentially the same as the item imputation rate) was typically much higher in the 1990 long form than in the 2000 ACS in Bronx County. Furthermore, this difference was strongly related to the lower quality of field data collection for census long-form information in comparison to the ACS. The U.S. General Accounting Office (2002a) reported on preliminary work carried out by the Census Bureau for long-form items in which the imputation rates were slightly higher than for the C2SS.

More complete work on comparing item imputation rates between the C2SS and the 2000 census long-form is reported in National Research Council (2004:Ch.7). Table 4-1 reproduces their analysis of comparative imputation rates for 15 person items and 12 housing items, by type of return. Imputation rates were lower for the C2SS than for the census long-form sample with one exception (year structure built). As National Research Council (2004:284–286) describe, C2SS enumerator (follow-up) returns required less imputations than mail returns for all person-item

Table 4-1 Imputation Rates for Selected Long-Form Items, 2000 Long-Form Sample and Census 2000 Supplemental Survey, by Type of Response, Household Population (weighted)

	2000 Long Form			Census 2000 Supplementary Survey		
Item	Total	Self	Enum-erator	Total	Self	Enum-erator
Person Items						
Marital Status	2.2	1.4	4.3	1.8	2.4	1.0
Educational Attainment	7.2	5.2	12.0	4.8	4.9	4.7
English-Speaking Ability	7.6	7.3	7.9	6.0	10.5	2.3
Place of Birth	9.2	7.8	12.5	6.4	8.1	4.1
Residence 5 Years Ago	5.8	4.3	9.6	4.0	5.6	1.8
Physical Activity Disability	7.6	7.1	8.9	5.2	7.4	2.1
Work Disability	11.4	12.2	9.3	5.9	8.3	2.2
Veteran Status	7.5	6.1	11.0	4.7	6.1	2.5
Employment Status Recode	11.1	10.2	13.4	6.0	8.2	2.6
Place of Work - State	9.7	7.3	15.5	5.8	6.5	4.8
Transportation to Work	7.6	5.4	13.0	4.6	5.5	3.3
Occupation Last Year	14.9	13.2	19.2	9.5	11.1	7.1
Weeks Worked[a]	19.3	18.6	20.9	9.6	11.1	7.3
Wage and Salary Income	20.0	15.0	32.6	16.4	13.0	21.4
Income, All Sources[a]						
100 Percent Imputed	24.5	18.9	38.5	20.0	16.1	25.7
Some Imputed[b]	29.7	25.5	40.3	23.9	20.7	28.6
Housing Items						
Units in Structure	4.4	4.9	3.0	1.4	1.6	1.0
Year Structure Built	11.7	9.3	18.0	13.4	7.4	22.8
Number of Rooms	6.2	6.2	6.4	2.6	3.4	1.4
Complete Plumbing	3.4	3.5	3.1	1.0	1.4	0.3
Complete Kitchen	3.4	3.5	3.1	0.9	1.3	0.3
Fuel Used for Heating	7.4	6.3	10.1	2.1	1.6	2.8
Electric Cost[c]	17.1	13.6	26.1	6.9	4.3	11.0
Monthly Rent	15.6	13.2	19.2	5.3	4.2	6.3
Property Taxes	32.0	27.0	49.6	20.8	13.7	35.4
Value of Property	13.3	12.3	16.6	9.7	6.0	17.4

NOTES: Rates (percents) exclude assignments. In 2000, self responses included mail, telephone, Internet, and Be Counted returns; enumerator responses included forms obtained in nonresponse follow-up, list/enumerate, and other field operations. In the C2SS, self responses included mail; enumerator responses included forms obtained in telephone and in-person follow-up.

[a] For 1999 in the 2000 census long-form sample; for last 12 months in the C2SS.
[b] Includes 100 percent of income imputed.
[c] Annual cost in the 2000 census long-form sample; last month's cost in the C2SS.

SOURCE: National Research Council (2004:Table 7.5); tabulations by U.S. Census Bureau staff from the 2000 Sample Census Edited File (SCEF) and the Census 2000 Supplementary Survey edited file, provided to the Panel to Review the 2000 Census spring 2003.

characteristics except income, while 2000 census long-form sample self (mail) returns tended to be more complete than enumerator returns for the same characteristics. This suggests the effect of better training among the C2SS interviewers relative to the temporary census enumerators. It also suggests the effect of the priority in the census of collecting *at least* (and, if necessary, only) the basic complete-count items like age and gender, whereas the C2SS interviewers are more likely to press for responses to all questions. However, Table 4-1 suggests that this distinction does not apply to the housing data items; for those questions, enumerator returns require more item imputation than self returns for both the C2SS and census long-form sample.

4–D.2 Quality of Imputed Responses

Rates of unit and item nonresponse are only partially informative as measures of the error rate due to nonresponse. This is because the imputation and weighting routines that the Census Bureau uses to treat item and unit nonresponse (and survey undercoverage) can offset some of the information loss, depending on the extent to which the various assumptions used to support the imputation methods hold (e.g., responses missing at random). Therefore, measures of the quality of imputations are an important additional measure of the impact of item and unit nonresponse.

This impact could be measured either through a reinterview survey or through matching to a more reliable source of data (possibly administrative records or highly reliable household surveys). But both approaches are problematic. Reinterview surveys of appreciable sample size are expensive and require high-quality interviewing to elicit higher-quality responses than provided earlier. Matching studies are limited by the availability of higher-quality, comparable information—a difficult standard to meet. The Census Bureau is conducting a matching study comparing C2SS responses to those for the 2000 census short form, though errors in both systems complicate the comparison.[7]

[7]Due to the design of the C2SS—specifically, the provision that the same respondent would not receive both the census long form and the C2SS—this matching is feasible only for characteristics on the census short form.

Some interesting work has been done on responses to race and ethnicity questions (Bennett and Griffin, 2002). A less satisfying variant of this analysis could still be carried out for small geographic aggregates—for example, comparing census and ACS frequencies and means for responses at the tract level, which would overcome the inability to match individual long-form responses. Some of this work is being conducted by the Census Bureau and is discussed below. Historically, there were matching studies of census responses to Internal Revenue Service (IRS) and Current Population Survey (CPS) data for earlier censuses (Bureau of the Census, 1964, 1975b),[8] and excellent reinterview studies were done in the 1970s and 1980s (Bureau of the Census, 1970, 1975a). Also, limited research on the quality of the imputations for 1990 was carried out by Thibaudeau (1998), but comparable work has not been done for 2000.

4–D.3 Measurement Error

Measurement error consists of differences between the response that was intended by the survey designers given a household's characteristics and the response that was actually captured. Possible contributors to measurement error include: misunderstanding of a question by the respondent, collecting data for the wrong time period, responding in the wrong units, transposing digits, making errors in capturing the response, intentional lying by either the respondent or the field enumerator, and so on.

It is reasonable to assume that, generally, the measurement error in ACS will be either comparable to, or very possibly somewhat less than, that for the census long form. This assumption follows from ACS design specifications: the ACS interviewing staff will be more experienced than short-term census enumerators and will be forbidden to use proxy respondents.

[8]Confidentiality concerns in the 1980s and 1990s led the IRS to restrict access to data, even for statistical purposes, thus precluding further census matching studies in recent decades. More recently, the IRS has facilitated limited administrative records research by the Census Bureau using IRS data with appropriate safeguards.

One challenge in comparing measurement error between the ACS and the census long form is reconciling the different definitions of residence in the two systems.[9] Both definitions are valid and defensible, and each may have particular advantages in different contexts, but their basic differences complicate comparison. Moreover, the ACS stages data collection over 3 months, and this may induce error due to temporary vacancies and frequent moving. For analytic purposes, the moving time window of the ACS may also present difficulties in interpreting quantities like income: each interview's snapshot is intended to capture a respondent's income for the 12 months preceding the interview, as opposed to the fixed prior calendar year reference frame of the census (1999, for income reporting in 2000), and this difference may complicate time-series comparisons.

4–E SUMMARY AND ASSESSMENT

4–E.1 Benefits and Costs

Arguably, the most compelling benefit of the ACS is the increased timeliness of its estimates relative to census long-form estimates: ACS data products are at most 3.5 years out of date when released while census long-form data products are never less than 2 to 2.5 years out of date and can be as much as 12.5 years out of date. Currently, using census data to develop lower bounds on the amount of year-to-year change for various estimates—for example, poverty rates—involves examining census-to-census differences and dividing by ten whereas this annual change can be measured directly under the ACS.

Operationally, relative to the decennial census, the prime advantage of a full-fledged ACS for the Census Bureau is the prospect of a short-form-only census. Though the census long

[9]The census attempts to capture "usual residence"—the location where respondents usually live or spend most of their time. By comparison, the ACS captures "current residence," the place where respondents are at the time of the interview. More precisely, the ACS uses a "Two Month Rule": any respondent at a sampled household unit who has been living at the location for more than two months is considered a current resident (U.S. Census Bureau, 2003c). This can create differences for migrant workers or "snowbird" retirees who live for lengthy periods in different areas of the country.

form is delivered to only a fraction of the population—in the 2000 census it was administered to approximately a 1-in-6 sample—the operational burden is tremendous. Completed long forms constitute a mountain of paper, and each form must be unstapled (running the risk of pages being mishandled) before processing. Moreover, the short form is inherently more manageable than the long form in terms of delivery options; though administration of a census long form via the Internet, a portable computing device, or telephone (interactive voice response) is no doubt possible, having to deal only with a short form makes implementation of these response technologies much simpler, not to mention more palatable to respondents.

A great strength of the ACS relative to other national household surveys is its large sample size, which allows it to provide small-area information about the American population, including population characteristics profiles for counties, cities, and other local areas. Over a 5-year period, the survey's sample size will approximate that of the census long form, supporting the production of estimates for small and nonstandard geographical areas, such as school districts and traffic analysis zones. In addition—and again given the large sample size—information will be available for population groups defined by factors other than geography, including racial and ethnic groups, age classes, occupational groups, and educational and health categories. (Tabulations can also be prepared for subpopulations with some combination of these characteristics.)

While the census long form can only provide these small-area profiles in once-per-decade snapshots, the ACS collects information continuously throughout the decade. Therefore, the ACS has the important advantage of providing estimates of the intercensal dynamics of small-area changes in the many variables listed above. Such estimates, which have been almost nonexistent up to now, can provide important information for policy initiatives and public and private planning.

The ACS may eventually permit researchers to develop an integrated framework for more accurate small-area estimation, perhaps combining one or more waves of ACS data with results from administrative records, other household surveys, and the short-form decennial census. This broader framework would incorporate the ACS as a supplement to the social and demographic information currently collected by existing surveys and administrative records systems. There are a variety of synergies that can be imagined between the ACS and household surveys such as the Current Population Survey, each drawing on the other to improve the information collected.

As we commented in detail in Section 4–D, much work must be done to clarify the quality and accuracy of ACS estimates from a statistical standpoint. That said, there are also reasonable arguments from the perspective of survey methodology that the ACS may provide more accurate information than the census long form. During ACS nonresponse follow-up operations, ACS data would be collected by interviewers with substantially more experience—having done the work continuously—than the temporary enumerators employed during the census. Moreover, by spreading the demand on respondents to provide detailed personal and household information over the decade, the ACS may also be less susceptible to flaws and inaccuracies that may arise from nonresponse in a once-a-decade measurement. During the 2000 census, concern over the perceived intrusiveness of the long-form questions was well publicized, leading to the conjecture—albeit one that has not been empirically documented—that this concern may have negatively affected response rates on long-form questions and, accordingly, impaired the accuracy of long-form data.

These benefits—particularly the key benefit of increased timeliness—must be weighed against the costs of the ACS. Given that it cannot "piggyback" on some of the infrastructure provided by the decennial census, one might assume that the ACS could cost more than the marginal cost of the long form it is replacing. However, the Census Bureau has argued that operational efficiencies will make a short-form-only census complemented by the ACS a less expensive option than a mixed long-and-short-form traditional census. In congressional testimony on May 13, 2003, Census Bureau director C. Louis Kincannon commented that "our current estimates indicate that three components of the 2010 Census [the ACS, the MAF/TIGER Enhancements Program, and early planning and testing] will cost approximately $11.2 billion. However, if we change course right now and revert to a traditional census, the cost will increase to more than $12 billion and perhaps much more."[10]

In its original presentation of its 2010 census strategy, the Census Bureau argued that most of the additional costs of the ACS can be paid for through the associated efficiency gains in the 2010 census. According to the Bureau, these savings would result by eliminating the collection and processing of long-form information during the decennial census, through improvement of MAF/TIGER, and through use of hand-held data collection devices to facilitate field follow-up of mail nonrespondents. As the panel noted in its letter report (National Research Council, 2001c), we have not seen validation of this claim based on empirical evidence and suggest that a fuller cost-benefit analysis of the ACS would help bolster the case for the survey.

In 1995, a previous Committee on National Statistics panel studying the decennial census offered its comments on an idea "which the Census Bureau has recently been investigating":

> to drop the long form from the census and substitute a continuous measurement survey—that is, a large monthly survey of perhaps 200,000 to 500,000 households. By averaging the results of the monthly surveys over a period of 3 to 5 years, more timely long-form-type data, accurate enough for use in relatively small geographic areas, could be produced. . . .
>
> In its preliminary work, the Census Bureau has speculated that the costs of the new continuous measurement survey over a decade could be roughly offset by the cost savings from dropping the long form from the census and by other cost reductions that might be achieved in intercensal operations. . . .
>
> Although we believe that the proposed continuous measurement system deserves serious evaluation, we conclude that much work remains to develop credible estimates of its net costs and to answer many other fundamental questions about data quality, the use of small-area estimates based on cumulated data, how continuous measurement could be integrated with existing household surveys, and its advantages compared with other means of providing more

[10]The remarks are quoted from the director's prepared testimony before the U.S. House Subcommittee on Technology, Information Policy, Intergovernmental Relations, and the Census at a hearing on the ACS' potential to replace the census long form in 2010.

> frequent small-area estimates. In our judgment, it will not be possible to complete this work in time to consider the use of continuous measurement in place of the long form for the 2000 census (National Research Council, 1995:9).

Nearly a decade later, faced with the task of offering advice on making the vision of continuous measurement a reality in time for the 2010 census, the similarity between the arguments then and now is uncanny. Similar, too, are the points of concern; the current panel is hard-pressed to improve upon the basic summary of concerns outlined by our predecessors. We are, however, much more sanguine that a compelling case can be made for the ACS and that it is a viable long-form replacement in the 2010 census.

In summary, the panel appreciates the enormous potential benefit of the ACS—of having a program for continuous measurement of key social and demographic variables of national interest. The ACS presents a unique source of timely information that could be extremely useful to public and private planning and that could be used to support more effective and targeted fund allocation. The potential benefits of the ACS are self-evident and require little salesmanship. However, what does require fuller justification is how these benefits offset the costs of the program and, more fundamentally, how the program works as a true long-form replacement. The panel is optimistic that such a compelling case can be made, though it will take continued evaluation work and research.

> ***Recommendation 4.1:*** **The Census Bureau should continue research to understand the differences between and relative quality of ACS estimates and long-form estimates, with particular attention to measurement differences and error from nonresponse and imputation. The Bureau must work on ways to effectively communicate and articulate those findings to interested stakeholders, particularly potential end users of the data.**

The fact that the Census Bureau has not done more in comparing the data collected from the thirty-one ACS test sites, the

C2SS, and the 2001 and 2002 Supplementary Surveys with the data collected by the 2000 census long form is disappointing.[11] Such analyses could be used to assess the quality of ACS data and would be helpful in making the argument for transition from the long form to the ACS. This deficiency is probably due to limited analytic resources at the Census Bureau and creates an argument for "farming out" this analysis to outside researchers. Furthermore, since access to local information is very useful in interpreting the results, the Census Bureau should explore whether local experts might be interested in assisting in this effort.

> ***Recommendation 4.2:*** **The Census Bureau must make ACS data available (protecting confidentiality) to analysts in the 31 ACS test sites to facilitate the comparison of ACS and census long-form estimates as a means of assessing the quality of ACS data as a replacement for census long-form data. Again, with appropriate safeguards, the Census Bureau should release ACS data to the broader research community for evaluation purposes.**

> ***Recommendation 4.3:*** **The Census Bureau must issue a guide for users of ACS data that details the statistical implications of the difference between point-in-time and moving average estimates for various uses.**

Part of a fuller justification of the ACS necessarily involves a cost-benefit assessment: enumeration of all benefits and costs, measurement or postulation of the benefits and costs, and comparison with costs and benefits (including data collection and processing) of the current approach (the census long form). The panel acknowledges that it is difficult to put a price tag on the value of more timely data, but coming to terms with cost-benefit

---

[11] As this report went into production, the Census Bureau released additional information on ACS quality metrics on their Web site at http://www.census.gov/acs/www/UseData/sse/index.htm [3/1/04]. In addition, the Bureau has commissioned some studies comparing 1999–2001 ACS and 2000 census long-form-sample data for several ACS test sites; those are expected to be released later in 2004.

trade-offs is an important part of assessing the program. Realistic assessment of the costs and benefits is complicated by the fact that so much remains to be demonstrated regarding the relative accuracy of ACS estimates and their applicability in the host of applications that currently use long-form data. We are optimistic that increased Census Bureau attention to informing data users and stakeholders (whether established users of the long-form data or newcomers) about the unique features and challenges of working with ACS data will build a stronger case for the survey.

4–E.2 ACS Funding

Given our panel's charge, the most basic question we face is whether the ACS is a satisfactory replacement for the census long form (and therefore something that should be the foundation of 2010 census planning as it has become). We recognize that significant estimation and weighting challenges must be addressed; the survey's costs, benefits, and uses must also be clearly articulated in order to convince users and stakeholders of its effectiveness. However, we do not see any looming flaw so large in magnitude that full ACS implementation should be set aside.

We therefore encourage full congressional funding of the ACS and are heartened that funds for launch of the full-scale ACS in late fiscal 2004 have been approved. Moreover, the administration's fiscal 2005 budget request covers a first full year of funding for the ACS. We emphasize that it is important for the continued role of the ACS in 2010 to be decided early, within the next 2 years. Implementation of the ACS would allow the 2010 census to consist only of the short-form questionnaire, a design feature that is too critical and too wide-reaching to leave unresolved until late in the decade. The short-form-only census would facilitate broader Internet data collection and the use of PCDs to collect respondent data; it would also reduce the data collection effort and simplify use of multilanguage forms. But a late reemergence of the need for long-form data collection would remove any efficiencies the Census Bureau had developed from its streamlined design.

> ***Recommendation 4.4:*** **As soon as possible, based on the 2006 proof-of-concept test, the Census Bureau should work with Congress and the administration to secure agreement on the overall design for the 2010 census and the American Community Survey (ACS). Extended delay in finalizing an overall design for the 2010 census—such as occurred in the preparation for the 2000 census—would unacceptably heighten the risk associated with the 2010 census. The role of the ACS is of particular concern; failure to secure commitment to the ACS as a replacement for the census long form would severely impair plans for a short-form-only census and undercut the ability to provide reliable small-area characteristics data by 2010.**
>
> **The Census Bureau should identify the costs and benefits of various approaches to collecting characteristics information if support for the full ACS is not forthcoming. These costs and benefits should be presented for review so that decisions on the ACS and its alternatives can be fully informed.**

Funding for the ACS is, of course, not a decision of the Census Bureau but of Congress. The panel is encouraged by statements in a recent hearing on the ACS that indicate that congressional authorizers are aware of the importance of making a clear and early decision about ACS funding. Specifically, at a May 13, 2003, hearing on the ACS, Representative Adam Putnam (R-FL), the chairman of the House Subcommittee on Technology, Information Policy, Intergovernmental Relations, and the Census, commented in his opening statement:

> I am also very aware that we are rapidly approaching the point where the Census Bureau needs to know one way or the other if there will be a long form in the 2010 census or will the ACS be the new survey tool. It's fundamental to a successful 2010 Census that we let the Census Bureau know as soon as possible how the Congress expects the Census to be conducted. I'm hopeful that we can continue to work together to resolve these final remaining issues, and that Congress can make a final determination on full funding for the ACS in the near future.

It is essential, however, that Congress recognize that funding of the ACS is a long-term commitment. The benefits of the ACS will be jeopardized if the survey program is faced with oscillating budget commitments. Cuts in funding (and with them reductions in sample size) will impair the overall quality of the survey, with the most pronounced impact on the ability to produce estimates for small geographic areas and population groups. We strongly encourage the Census Bureau to conduct research that quantifies the sensitivity of ACS-based estimates to fluctuations in sample size, in order to make the case for sustained ACS funding more compelling to policy makers.

4–E.3 Contingency Planning

We endorse the ACS and strongly recommend that it replace the long form in the 2010 census. That said, we must reiterate our recommendation from previous reports that the Census Bureau begin contingency planning to be prepared should support for the ACS not be forthcoming. In our letter report (National Research Council, 2001c), we strongly urged the Census Bureau to make contingency planning a focus of its planning efforts, with particular attention to the funding levels for the ACS. The difficulty of securing fiscal 2003 funding for the full launch of the ACS underscored the importance of that recommendation. The obvious fallback contingency plan is reinstitution of the census long form; however, the costs and benefits of other options—such as implementation of a one-year ACS, operating simultaneously but not bundled with the census just as the C2SS operated in 2000—need to be developed and presented for review so that decisions on the ACS can be fully informed.[12]

> ***Recommendation 4.5:*** **The Census Bureau should identify the costs and benefits of various approaches**

[12]The Office of Inspector General of the Census Bureau's parent agency, the U.S. Department of Commerce, has expressed similar concerns. "If the Bureau does not receive sustained ACS funding throughout the decade, it may be unable to eliminate the long form for 2010"; consequently, the Census Bureau's planning for 2010 should "include a contingency plan for use of the long form" (U.S. Department of Commerce, Office of Inspector General, 2002:iv).

to collecting characteristics information should funding for the full ACS not be forthcoming. These costs and benefits should be presented for review so that decisions on the ACS and its alternatives can be fully informed.

4–F TOPICS FOR FURTHER RESEARCH AND DESIGN CONSIDERATION

A substantial agenda of outstanding operational and methodological issues should be addressed in order to ensure a fully operational ACS. Some of these issues should be tackled in the near future in order to generate the maximum benefits from use of the ACS as part of an integrated framework of estimates.

In addition to the research and design issues we raise here, other issues are described in other sections of this report. In particular, reconciliation of the census and ACS definitions of what constitutes residence at a particular location deserves prompt consideration (Section 5–B.3). Likewise, the effects on response of the mode in which the ACS is administered (Section 5–D.2) merit further examination.

4–F.1 Group Quarters

The intent of the census long form is to provide information on characteristics of the entire population. This means not only the population residing in housing units but also those living in group quarters, such as college dormitories, military barracks, prisons, and medical and nursing facilities. Nonresponse to the census long form and the need to impute for nonresponse may detract somewhat from the overall reliability of census long-form data, but those data do at least allow users to make some inferences about the group quarters population. Accordingly, the complete elimination of the census long form—and the possible loss of data on the group quarters population—is an obvious concern of some census stakeholders.

In its draft operational plan, the Census Bureau has indicated that the ACS will be administered to a 2.5 percent sample each year from the Bureau's group quarters roster (U.S. Census Bureau, 2003c). It remains to be determined how adequate this may be for monitoring this important population group, especially for small geographic areas and small demographic population groups. In Section 5–B.2, we recommend a complete reexamination of the Census Bureau's approach to enumerating the group quarters population. Continuing research and planning to ensure that this population is adequately covered in the ACS would not only contribute to a better enumeration but also bolster the case for the ACS' unique role relative to other federal household surveys.

4–F.2 Voluntary versus Mandatory Response

The law governing conduct of the census imposes penalties on "whoever, being over eighteen years of age, refuses or willfully neglects ... to answer, to the best of his knowledge, any of the questions on any schedule submitted to him in connection with any census or survey" enabled in other parts of the census code (13 USC § 241(a)).[13] In addition, it is a crime to willingly give false answers to such censuses or surveys (13 USC § 241(b)). Accordingly, census mailings in 2000, as in previous years, prominently featured notices that "your response is required by law."

The Census Bureau has argued that because the ACS is intended to replace the mandatory census long form it should be conducted on the same mandatory basis as the census. The General Accounting Office has concurred that the Bureau has statutory authority to conduct the ACS and to require responses (U.S. General Accounting Office, 2002b). The distinction between voluntary and mandatory completion is significant because it is believed that the words "required by law" on the census forms are effective in raising response rates.

However, early congressional discussion of the nature and content of the ACS led individual members of Congress to suggest that the ACS be conducted on a voluntary basis. Accordingly, the Census Bureau conducted part of the 2003 Supplementary Survey (the prototype ACS) on a voluntary basis; this test included replacing the phrase "required by law" with a more

[13]However, the census code does provide that respondents cannot be compelled to disclose their religious beliefs or affiliation (13 USC § 241(c)).

generic appeal (U.S. Census Bureau, 2003c). [Rather than alter the instruments and scripts used in telephone or personal visit follow-up on a case-by-case basis, the Census Bureau conducted both types of follow-up using the voluntary participation language.] The response rates, including item nonresponse rates, on the voluntary surveys were compared with results from those obtained one year earlier in the 2002 Supplementary Survey.

Preliminary test results were publicly released by the Census Bureau in December 2003 (U.S. Census Bureau, 2003f); a fuller report and analysis is indicated as pending. The Bureau found that mail response dropped by over 20 percentage points when response was changed from mandatory to voluntary; based on the decline, the Bureau projects that a voluntary ACS would increase the annual cost of the survey by at least $59.2 million (U.S. Census Bureau, 2003f:vi). The Bureau also found evidence that participation in a voluntary-response survey was worse in areas that had low mail response to the 2000 census, leading the Bureau to conclude that voluntary methods might "compromise [its] ability to produce reliable data for these areas and for small population groups such as Blacks, Hispanics, Asians, American Indians, and Alaska Natives" (U.S. Census Bureau, 2003f:vi). If respondents decided to fill out the questionnaire, the survey results indicated that the voluntary designation did not degrade responses to individual items; voluntary and mandatory methods generally resulted in comparable levels of item nonresponse.

The mandatory versus voluntary distinction is an important one to resolve. The Census Bureau should continue work to assess the impact on nonresponse follow-up costs based on the change (likely, a decrease) in mail response if the full ACS is labeled voluntary rather than mandatory.

4–F.3 ACS as Both a Census Process and a Federal Survey

When fully implemented, the ACS will occupy a unique niche among the statistical data series collected by the federal government. Because it is intended to replace the census long form, the ACS should properly be viewed as a parallel component of the census process. It will be charged with producing the small-area and small-demographic-group data required for many legal and regulatory purposes and used in many research applications. In its sweep, the ACS will require development of a technical infrastructure on par with that for the decennial census itself (see Chapter 6). Pending resolution of the debate described in the previous section, it may also bear the notice that responses to the survey are required under the law. All that said, the ACS could also be properly viewed as one of many surveys fielded by the federal government on a number of topics. This dual role of the ACS—census component and federal survey—raises concerns that will require attention in coming years.

Primary among these concerns is the substantive overlap between the ACS and other federal surveys such as the Survey of Income and Program Participation, the American Housing Survey, and—especially—the Current Population Survey (CPS). As the Census Bureau works with Congress to secure ACS funding, the panel recognizes that it is virtually inevitable that the question will be asked as to whether other surveys might be cut back or eliminated to help pay for the ACS (or vice versa). In our assessment, a fully operational ACS is not immediately exchangeable with other surveys. For instance, as the potential basis for an estimate of the poverty rate, the ACS has the advantage of larger sample size but does not cover socioeconomic and poverty-specific questions with the same depth as the CPS. The CPS has the further advantage of years of experience in soliciting detailed economic information; face-to-face interviewers acquire fuller knowledge of the survey content area and may be able to assist CPS respondents in interpreting survey questions in ways that the broader-focus ACS interviewers may not be able to match. It is decidedly premature to offer any sort of guidance on whether the ACS or another federal survey should be preferred in given situations. The panel suggests further evaluation and exploration of relative data quality in topic areas where the ACS overlaps with other federal surveys. Research should also consider ways in which the ACS could support or supplement other federal surveys, including possibilities for using recently-collected ACS characteristics data to refine the sampling frames from which other surveys are drawn (for instance, targeting surveys to low- or high-income areas).

The second major concern regarding the dual role of the ACS is how the ACS will be treated within the census hierarchy. For cost savings, the ACS and short-form-only census plans should be coordinated in order to avoid redundant effort and to "piggyback" on existing structures when possible (e.g., to perform data capture using the same optical character recognition technology and equipment). Since the panel issued its second interim report, the Census Bureau transferred ACS authority and activities from its Demographic Programs directorate to the Decennial Census directorate, the same division that plans and operates the decennial census process (see Box 2.2). The full implications of this organizational move remain to be seen. The panel suggests, however, that the Census Bureau not lose sight of the inherent sample-survey nature of the ACS. While its weighting, editing, and imputation techniques may be similar to to those used in census operations (and, in particular, to past long-form implementations), they should also differ when appropriate and not be constrained to treat census and ACS returns in the exact same manner. It may also be useful, in the future, for the ACS to leave open the possibility for experimental components such as occasionally occur in federal surveys. These experimental components could include one-shot (or periodic) modules of questions on particular topics such as crime victimization or health care or on items of interest to a particular state or region. Experimental components might also include more general tests of proposed survey practices, such as was done in the test of voluntary versus mandatory response.

4–F.4 Revisiting Sampling Strategies

The basic ACS sampling strategy is simple: each month a systematic sample of approximately $\frac{1}{480}$ of the addresses on the Master Address File is taken, with one-third of mail and telephone nonrespondents randomly chosen for in-person follow-up. A number of variations on this basic strategy are either currently designed or under consideration for later implementation in the ACS by the Census Bureau. These additional possibilities are: (1) oversampling of governmental units with small populations, such as small towns, (2) oversampling of minority areas, and

(3) differential sampling of areas with poor initial mail and telephone response. We briefly comment on each of these possibilities and some additional methods not currently contemplated for implementation.

With respect to oversampling of small areas, the Census Bureau intends to use some version of the decennial census long-form design. In the 2000 census, sampling rates were 1 in 2 for governmental areas (counties, towns, townships, and school districts) with fewer than 800 occupied housing units (fewer than about 2,100 people); 1 in 4 for governmental areas with 800–1,200 occupied housing units (about 2,100–3,100 people); 1 in 6 for census tracts with fewer than 2,000 occupied housing units (fewer than about 5,200 people); and 1 in 8 for larger census tracts. The justification for this plan in the decennial census was originally to support reliable estimates of per capita income for small governmental units for use in fund allocation as part of general revenue sharing. However, this oversampling has been retained past the elimination of general revenue sharing because it tends to make coefficients of variation more equal across areas with different population sizes. Undoubtedly, that is the current justification for oversampling in the ACS. However, a new set of sampling rates may serve that purpose more effectively, and therefore, after the ACS has been in operation for a short while, it would be useful to compute the coefficients of variation for all responses on the ACS questionnaire for areas with different population sizes, to determine whether a different strategy might prove to be superior with respect to this objective.

For oversampling of minority areas, the Census Bureau has mentioned an interest in increasing the ACS sampling rate in areas with a high percentage of minority residents in order to provide estimates with lower coefficients of variation for important statistics historically related to racial and ethnic disparities. While the panel believes that this is justifiable, it should be understood that the historically lower mail return rates for minority populations could result in additional nonresponse follow-up costs for the ACS.

In terms of the differential sampling of areas with poor initial mail and telephone response, it is true that without it these areas will have much larger coefficients of variation than areas with high mail and telephone rates. Therefore, efforts to balance these coefficients of variation are justified.

Clearly, the ACS could be modified in many ways to better satisfy various purposes. While most of the above possibilities have various disadvantages that might argue against their implementation, it would be very helpful for the Census Bureau to provide arguments to help justify the current design. These are topics that need little or no additional data collection or field work to further develop. Rather, what is needed is summary information that is already available from the ACS in the test sites. We encourage the Census Bureau to provide some analysis along these lines.

4–F.5 Interaction with Intercensal Population Estimates and Demographic Analysis Programs

One high-priority research area should be the development of models that combine information from other sources—such as household surveys, administrative records, census data, and the like—with ACS information. One prominent example of this is the interplay of ACS estimates and the Census Bureau's population estimates program. At this point, it is planned that estimates from the ACS are to be controlled to postcensal population estimates at the county level and some degree of demographic aggregation. However, this should not be considered a one-way street. It is also possible for the ACS to be used to provide the population estimates program with improved estimates of internal and external migration, fertility, household size, and vacancy status. The resulting improved population estimates could then be used as improved marginal totals to which to control ACS estimates. Because the ACS also provides direct information on population size, a joint estimate from population estimates and from the ACS is conceivable. The Census Bureau should (1) conduct research on how the ACS can be used to improve intercensal population estimates, and (2) examine how existing household surveys could change their poststratification practices (controlling totals by age, race, and sex) given the collection of ACS data.

The potential for the ACS to provide improved estimates of internal and external migration also suggests the importance of exploring possible interactions between the ACS and population estimates derived by demographic analysis. Demographic analysis uses aggregate data on birth, death, immigration, and emigration to produce population estimates by age, sex, and race. It was a key benchmark used to evaluate coverage in the 2000 census, but it has significant limitations. First, estimates of immigration and emigration—particularly those of undocumented immigration—are inherently difficult to produce with precision. Second, existing administrative records used to generate demographic analysis counts facilitate only the most basic racial comparisons—white and black—but do not permit direct estimation of Hispanics and other groups. The Census Bureau should consider ways in which the ACS might inform demographic analysis estimates, including more refined estimators of the size of the foreign-born population and of internal migration. We discuss further possible improvements for demographic analysis in 2010 in Section 7–B.[14]

Other possibilities—for instance, using ACS and household survey information jointly in regression models to provide improved estimates of the frequency of crime or unemployment—could also be fruitfully addressed as a research topic.[15] Another high-priority research area should be identification of better procedures for weighting and imputation, to address nonresponse and undercoverage in the ACS; the hope would be to develop procedures that are, in a sense, optimized for ACS survey data, and not simply borrowed from procedures used on the decennial census long form.

---

[14]The methods by which the ACS data could be used to improve demographic analysis could also be applicable to improvements of intercensal population estimates for the nation as a whole (National Research Council, 2000b; Citro, 2000).

[15]The use of models that combine information from other sources has implications for the sample designs of the major household surveys and is a future research topic of great potential interest. Use of these models and connections to external programs such as the ACS may permit other household surveys to reallocate sample to areas in which estimates are less reliable.

4–F.6 Research on General Estimation Issues

The challenges of implementing the data collection for the ACS have understandably been given the highest priority at the Census Bureau. As a result, relatively straightforward estimation methods have been proposed for use in the short term, deferring estimation improvements for later. Unfortunately, this has meant that little research has been done on alternative approaches to estimation. We mention here some issues that should be examined by the Census Bureau once data collection is under control:

- *Alternatives to moving averages.* Moving averages are easy to implement and have well-understood properties, including variance reduction. However, they will reduce large deviations that obtain for shorter periods of time than the smoothing window. There are methods for reducing this feature of moving averages that still retain much of the variance reduction benefit.[16]
- *Controlling versus combination.* Current plans are to control ACS population estimates at the county and major demographic group level to postcensal population estimates. For initial implementation, this is a reasonable approach to take, since it will likely improve the quality of the ACS population estimates. However, the use of the ACS in combination with information from various data sources—including census data, data from household surveys, and data from administrative records—needs to be a two-way street, as the ACS will provide independent information on population size and various characteristics information formerly obtained from the long form. Specifically with respect to population size, the ACS will produce estimates at the county and major demographic group level that will have relatively large variances for most smaller counties, but because they are independent, they could still be used to improve postcensal population estimates. This will be more certain the further one moves away from a census year, as postcensal population estimates are increasingly

[16]Two possibilities that could be examined are state-space time-series models and spline smoothers.

variable as one moves further into each decade. In addition, the ACS data on population would not need to be used directly. Instead, data from the ACS could be applied to components of the postcensal estimates program, in particular estimates of interstate mobility, fertility, and household occupancy. Finally, there is the much more demanding vision of the ACS underlying a small-area estimates program, whereby information from the above sources is used in conjunction with the ACS to produce a wide variety of small-area information of higher quality than could be provided by any individual data source. Given the varying quality of data from ACS and other sources, ACS data should not simply be controlled to data from these other sources; instead, hierarchical models should be used that will let the data from the various sources determine the degree to which estimates are combined. This latter vision in totality is certainly beyond the current research literature in terms of complexity of application, especially since many of the proposed data sources might be inconsistent. However, initial efforts should be undertaken since the methods to carry out simple versions of this possibility currently exist and are regularly used in other applications.

- *Weighting and imputation methods.* The Census Bureau currently intends to use ten or so different weighting methods to accommodate: (1) the sample design of the ACS, (2) the use of data from different months (and modes) of response to compute the estimates for a given month and area, (3) whole household nonresponse, (4) individual unit nonresponse, (5) individual item nonresponse, and (6) undercoverage. These methods were adopted because of their current use (when relevant) in processing the decennial census short and long forms, and because of their resulting recognized benefits and ease of implementation in that very similar setting. Some of these weighting approaches are entirely appropriate for the ACS, and some are unique to the ACS as they are meant to address differential mode effects and the more complex sample design of the ACS relative to the long form. However, the current use of sequential hot-deck imputation for the treatment of individual item nonresponse, and the use of variance estimates that ignore the contribution of item nonresponse, are methods that are no longer representative of the current state of the art. Furthermore, it is not clear that nonresponse and undercoverage for the ACS will be sufficiently similar to these problems for the long form that these various long-form weighting methods should be utilized in the ACS without additional supporting research. The particular problem of the treatment of item nonresponse is becoming increasingly important given the degree of nonresponse experienced in the 2000 census.

CHAPTER 5

Enumeration and Data-Processing Methods

BUSINESSES WILL NOT TAKE A HOLIDAY, travelers will not cut short their trips, citizens will not simply stand still and stay put—in short, the collective life of the nation will not take pause on April 1, 2010, simply because it will be Census Day. Nor has the population taken pains to make itself easy to count on any previous Census Day. Instead, when 2010 arrives, the Census Bureau will confront what it has always faced: an increasingly dynamic and diverse population, in which each person and household varies in both willingness and ability to be enumerated in the census. As a result, census-taking involves a continual search for methods to maximize participation in order to collect information on as many willing respondents as possible. It also involves the need for strategies to do everything possible to count those whose economic and living circumstances make it difficult for them to be enumerated by standard means.

For 2010, the Census Bureau proposes to make a significant change to its tool kit of enumeration methods. Relying on a short-form-only census and improvements to its geographic resources, the Bureau hopes to make use of a new generation of portable computing devices (PCDs) to enhance both the ease and accuracy of interactions between census enumerators and follow-up respondents. These devices have great potential to improve census-taking, but their development process involves considerable risk and uncertainty (Section 5–A). In Section 5–B, we comment on several areas—including group quarters and residence rules—where both the 2000 census experience and continually changing societal influences suggest the need for redefinition and recalibration of census-taking approaches. We then turn in Section 5–C to two particular enumeration challenges—representing different extremes of urbanicity—that we believe deserve attention in 2010 census planning. Section 5–D comments on the Census Bureau's plans to expand the means by which respondents can return their census information, including wider use of the Internet and telephone. Finally, the 2000 census experience focused attention on two processes—unduplication of person records and imputation for nonresponse—that are more commonly thought of as late-stage, data-processing functions. However, lessons learned from the 2000 census coupled with new technology will make these processes—and management of the critical trade-offs in cost and accuracy associated with them—a fundamental part of the enumeration strategy of the 2010 census (Section 5–E).

As we discuss PCD plans, it is important to note that the term "portable computing device" (PCDs) is the panel's, and not currently the Census Bureau's, usage. The Census Bureau uses the term "mobile computing device" or, more frequently, simply "MCD" to refer to the computers. However, the choice of MCD as a label is unfortunate because it conflicts with the abbreviation for "minor civil division," a long-standing concept of census geography that refers to the subcounty (township) divisions that are functioning governmental units in several Midwestern and Northeastern states. For this report, we have adopted the compromise label of PCD.

5–A PORTABLE COMPUTING DEVICES

Since the 1970s, computer-assisted interviewing (CAI) has emerged as a major element of modern survey methodology. Development began with computer-assisted telephone interviewing

(CATI), allowing interviewers to administer a questionnaire via telephone and capture responses in electronic form. In the late 1980s, the emergence of portable laptop computers initiated a second wave of development, computer-assisted personal interviewing (CAPI), in which interviewers conduct face-to-face interviews with respondents using a computerized version of the questionnaire on the laptop. The automated data capture that follows from CAI methods, along with the capacity to tailor questionnaires to individual respondents through "skip" sequences jumping to different parts of the questionnaire or customizing question text based on information already collected in the interview, have proven enormously advantageous, even though cost savings have proven elusive (see National Research Council, 2003b). In the 2000 census, field staff used laptop computers to collect data as part of the Accuracy and Coverage Evaluation Program. Hence, it is natural that plans for 2010 revisit—and try to improve—CAI implementation in the census.

During the past decade, a class of miniature electronic devices has entered the marketplace and continued to mature—the handheld computers commonly known as personal digital assistants (PDAs); they are also sometimes called pen-based computers, since the principal means of interacting with many of the devices is through handwriting on the screen.[1] Today, most of the devices use the Palm operating system or a Pocket PC (Windows CE) version of the Microsoft Windows operating system. As the technology continues to mature (and to get cheaper, faster, and more powerful), survey organizations have increasingly tested the potential use of these devices for their work. More recently, tablet computers (roughly, a hybrid device with the computing power and screen size of a laptop computer but in a one-piece, keyboardless design using handwriting recognition) and "smart phones" (combining PDA and cellular phone functions) have arrived on the market, and survey organizations have considered those devices, too, in the quest to outfit interviewers with easily portable but survey-capable computers.

[1]As text pagers and handheld e-mail devices have become more common, several newer PDAs use a miniature keypad in lieu of a pen/stylus and handwriting recognition.

For the 2010 census, the Census Bureau proposes to use portable computing devices "that will enable enumerators to locate and update address information for housing units, conduct interviews, transmit the data directly to headquarters for processing, and receive regularly updated field assignments" (Angueira, 2003b:3). In doing so, the Bureau hopes to exploit another increasingly common technology—global positioning system (GPS) receivers that can be embedded in portable computers. The devices that the Census Bureau is testing are of the Pocket PC class and, though no decision has been formally made, the information available to the panel through discussions with Bureau staff suggests that the current vision for the device in 2010 is of the same size as the current Palm/Pocket PC models.

In the panel's earliest discussions with the Census Bureau about the prospects of PCD use, the principal arguments raised by the Bureau in support of the plan were: that PCDs would save field costs by helping field staff complete their work more efficiently and without getting lost, that PCDs would save costs on paper and forms, and that PCDs would achieve familiar CAPI benefits such as automated data capture. In more recent interactions, though, the savings-through-better-navigation argument has been downplayed while much more emphasis has been put on savings on paper costs. Indeed, the Bureau's draft baseline design for 2010 maintains that "through the use of automation and [PCDs], we will significantly reduce the amount of paper used in the field (including questionnaires, address registers, maps, assignment sheets, and payroll forms) and the large number of staff and amount of office space required to handle that paper" (Angueira, 2003b:3).

5–A.1 Testing PCDs: Pretests and the 2004 Census Test

As portable computing devices began to emerge as a focus of the 2010 census plan, the Census Bureau initiated small pilot tests involving basic skills. For instance, small numbers of field staff with different levels of local familiarity were assigned to find a set of addresses using TIGER-based maps on a Pocket PC-class device in a pilot test in Gloucester County, Virginia. This test concentrated only on locating features using a small-screen map and not on using the computer to calculate a route to those features. In addition, the devices used in the test were not equipped with GPS receivers, so the test was not meant to assess the locational accuracy of the designated addresses (U.S. Census Bureau, Mobile Computing Device Working Group, 2002).

The 2004 census test is intended to serve as the first major proving ground for portable computing device usage in the census, and to that end is more comprehensive than the earlier pilot tests. A Pocket PC-class device equipped with a GPS receiver has been selected for the test. In addition to continuing to gauge enumerator reaction and ability to use the devices with a short amount of training, the primary thrust of the test is to assess the performance of a basic workflow for the devices. In 2000 and previous censuses, assignment of enumerators' workloads was quite hierarchical in nature, filtering from local offices down to crew leaders and finally to the enumerators. The workflow being used in the 2004 test centralizes control to a great degree at Census Bureau headquarters. Though local census office and crew leader input is sought in making initial assignments and elsewhere in the process, all information is channeled directly through headquarters; each enumerator's PCD communicates with headquarters in order to receive new assignments. Likewise, hierarchical processing when completed questionnaires are gathered is also replaced. Rather than having completed questionnaires undergo visual inspection by enumerator crew leaders and by local census office staff, the PCD transmits each enumerator's completed questionnaires directly to headquarters and on to data processing.

Members of the panel saw a demonstration of the device to be used in the 2004 test, and our understanding is that software development is still under way. At this time, the devices are able to provide enumerators with listings of their workload assignments and with maps of the blocks they will be working in, but there is no connection between the two. That is, enumerators cannot highlight one or more of their assigned cases, request that they be plotted on a map, and thus decide on an optimal route. In its current form, the map information appears to serve only a purely reference purpose. The 2004 test implementation is also preliminary in nature because paper—and not electronic transmission—will be used for many of the progress reports and initial assignment rosters that will circulate between census headquarters, local census offices, and crew leaders.

5–A.2 Assessment

The Bureau's plans for portable computing devices are a particularly exciting part of a reengineered census, but we are concerned about various aspects of their development, among them the following. First, much remains to be done to bolster the argument that use of PCDs will create major cost savings. Second, we are concerned that the Bureau's current approach to testing the devices may be based primarily on devices currently available on the market. Hemmed in by that limitation, the Bureau runs the risk of not placing enough emphasis on the establishment of requirements and specifications for these devices and of not adequately accounting for human factors.

PCD Cost Savings

PCDs are critical to the Census Bureau's plans to achieve cost savings in the 2010 census plans. Indeed, Bureau staff at the panel's last public meeting in September 2003 identified PCDs as the centerpiece of savings for the short-form data collection in 2010. The basic claim is that huge costs associated with current census field operations can be directly linked to the use of paper, including the cost of rental space to store the paper, the cost of the paper and printing itself, and the cost of distribution, transportation, and handling of the paper. The expectation is that the use of PCDs will produce sufficient savings through the reduction of paper to pay for itself. In addition to the savings associated with the reduction in paper, Census Bureau cost documents have asserted that PCDs will reduce equipment and staff needed in local census offices to produce maps, reduce costs of data capture, and improve productivity (U.S. Census Bureau, 2001a).

The problem is that, at present, the panel knows of no empirical evidence for any of these potential cost savings. Therefore it appears that the Bureau is proposing to spend a large amount of money in PCD procurement in the *hope* that the efficiency and paper-reduction gains will be achieved. This is too radical a departure from current procedures and too risky an undertaking for the decision to be made without careful testing and the accumulation of evidence that such cost savings could realistically be achieved, or that the use of the PCDs will pay for themselves without negatively affecting data quality (or alternatively, that the expected gains in quality offset the additional costs).

For example, much has been made of the use of global positioning system (GPS) receivers attached to these devices to reduce duplication through correct map spotting of dwelling units. To support this contention, we have heard many anecdotes about enumerators who cannot read maps, but have been shown no hard evidence of the extent of the problem. (It is also far from clear that the same enumerators who experience difficulty working from and reading paper maps will, with minimal training, be able to use the maps on a handheld computer with any greater efficiency or any less error.) Moreover, the use of GPS will do little to solve the problem of correctly identifying individual units within multiunit structures. The device may be able to indicate when the enumerator has reached the correct structure, but readings will likely be impossible in the structure's interior. It is also not clear that GPS receivers will be of much utility in high-density urban areas, where the map spot associated with an address may be based on interpolating from address ranges associated with streetline segments, which may not match precise structure locations. Since it is difficult to fully articulate or confirm the benefit of having a GPS receiver built into every PCD, it is even more difficult to contrast that benefit with the associated cost and decide if the expense is justifiable.

This is not to argue against technological advancement in conducting the census. What we do argue for is better explication of the costs and benefits; the Census Bureau's experience with converting its survey programs to computer-assisted interviewing methods amply demonstrates that new technology does not automatically translate to cost savings. The introduction of laptops for ongoing survey data collection increased the number of staff at Census Bureau headquarters, while not reducing (initially, at least) the number of processing staff at the Bureau's national processing facility in Jeffersonville, Indiana. No evidence has been forthcoming on how much the transition cost the Bureau, but the arrival of computer-assisted interviewing does not seem to have saved the Bureau any money (National Research Council, 2003b).

The decennial census experience with laptop computers for Accuracy and Coverage Evaluation (ACE) interviewing also suggests difficulty in assuming automatic cost savings with the implementation of technology. Titan Systems Corporation (2002) discusses the procurement of approximately 9,700 laptops for ACE from a vendor the Bureau had worked with since 1996. There were procurement and delivery problems, and a 10 percent overage was needed for failures. Specifically, the report notes that "the 9,639 laptop kits had to be assembled before shipping and this required the contractor to make BIOS configuration settings, load the software, and bundle the various accessories (adapters, manuals, batteries, etc.). The contractor had problems ensuring that each unit was configured as required" (Titan Systems Corporation, 2002:9). If the procurement of 9,700 laptop computers occasioned such problems, there seems to be legitimate concern about the procurement process for (as we understand it) some 500,000 PCDs.

The 2004 and 2006 census tests will be critical to establishing the veracity of the cost-saving assumptions associated with PCDs. However, we are not confident that the 2004 census test is capable of providing the basic data needed to make return-on-investment decisions for 2010. That test is posed more as a wide-ranging but ultimately tentative first-use test to establish basic feasibility. Accordingly, as we note throughout this report and particularly in Chapter 9, the onus is that much greater on the 2006 test as a proof of concept. The Bureau must build into that test appropriate measures and metrics to make a cost-quality assessment of the effectiveness of PCDs, and these measures need to include a realistic assessment of training costs, failure rates and associated maintenance and support costs, accuracy rates, efficiency improvements, and so on.

Testing, Requirements, and Human Factors

It would be a mistake to make assumptions at an early stage that unnecessarily limit the functionality or constrain the human factors of these devices. Given the rate of technological development, it is not unreasonable that a tablet-size PCD with a full-blown operating system, adequate memory, a 20-gigabyte hard drive, a GPS receiver, a modem, encryption facilities, and an 8-inch full-color screen display will be available in the market by 2007 at a price of $500 or less in the quantities required by the Bureau. So to prototype systems and to put too much emphasis on usability tests using devices of considerably less capability—rather than using early testing to further refine the basic logical and informational requirements that the final device must satisfy—is probably too conservative and will result in the acquisition and use of devices that will be less effective than necessary.

We strongly suggest therefore that the Census Bureau not focus on the particular limitations and capabilities of the existing 2 or 2.5-inch screen devices currently available on the market. In terms of the capability of the devices likely to be available for 2010, it is almost certain that some testing using high-end devices (e.g., tablet PCs) would provide a more realistic test.

The Bureau's most pressing need regarding PCD development is the definition of specifications and requirements—clear statements of exactly what the devices are intended to do. In Section 6–B.3, we suggest the designation of a subsystem architect with responsibility for PCD and field systems to address this need. A key part of establishing the specifications and requirements for the devices will be articulation of the other census operations besides nonresponse follow-up for which the devices may be used; it is unclear, for instance, to what extent PCDs might be used in American Community Survey operations or in block canvassing.

As the Bureau further develops its plans for PCDs, it will be essential to keep human factors in mind. The utility of the devices will depend on their effective use by a corps of temporary workers with relatively little training. While smaller devices of the current Palm/Pocket PC class may have advantages in terms of sheer size or weight, it is quite possible that working with a tablet-sized device will be much easier for census workers than repetitive pecking at a 2.5-inch screen. It is very important that the application software developed for the PCDs be tested by end users for its usability and accessibility, in addition to testing for computational bugs and flaws.

Recommendation 5.1: **The Census Bureau should develop and perform a rigorous test of its plans for use of portable computing devices, and this test should compare the performance and outcomes of data collection using:**

- **devices of the current (Pocket PC) class being developed for use in the 2004 census test;**
- **high-end devices (e.g., tablet computers) of classes that are very likely to be available at reasonable cost by the time of procurement for 2010; and**
- **traditional paper instruments.**

Such a test is intended to provide fuller information about the costs and benefits of portable computing devices, using paper as a point of comparison. The test should also provide the opportunity to review specifications and requirements for the PCDs, using devices of the caliber likely to be available by 2010.

Recommendation 5.2: **By the end of 2004, the Census Bureau should complete requirements design for its portable computing devices, building from the results of the 2004 census test and in anticipation of the 2006 proof-of-concept test. The requirements and specifications for portable computing devices must include full integration with the census system architecture and should include suitability for other, related Census Bureau applications. The Bureau's requirements design for PCDs must devote particular attention to the human factors underlying use of the devices.**

Recommendation 5.3: **The Census Bureau must develop a complete engineering and testing plan for the software components of the portable computing devices, with particular attention to the computer-assisted personal interviewing interface, data capture systems, and communication/synchronization capabilities (including assignment of enumerator workload).**

5–B CHALLENGING DEFINITIONS FOR A MODERN CENSUS

While PCDs offer the potential to improve the mechanics of census-taking, the panel believes that it is essential that 2010 census planners also take the opportunity for reexamination and change of some of the basic definitional concepts of the census.

5–B.1 Housing Units

First, and consistent with our recommendations in Chapter 3, the very notion of what constitutes a housing unit deserves a fresh assessment. It is largely for this reason that we recommend the creation of a Master Address File (MAF) coordinator position within the Census Bureau.

For census purposes, the MAF's most fundamental purpose should be to serve as a complete register of housing units. Accordingly, an important step in enhancing the MAF is an examination of the definition, identification, and systematic coding of housing units (and, by extension, group quarters). (See Sections 3–E.1 and 3–E.2 for additional discussion about housing unit identification and coding.) The current MAF/TIGER Enhancements Program may impart some benefit to MAF entries by virtue of their linkage to TIGER but does little to address two fundamental problems that hindered the MAF's effectiveness as a housing unit roster in the 2000 census.

The first of these is multiunit structures—physical buildings that contain more than one housing unit. A realigned TIGER database may offer a precise location for a structure—an aerial photograph may confirm a structure's existence or point to the construction of a new one—but that added precision is ultimately of little use if the address roster of subunits within the structure is unknown or inaccurate. Multiunit structures pose problems both conceptually (e.g., if the finished basement of a house is sometimes offered for rent, should it be counted as a unit?) and technically (e.g., do different data sources code an apartment as 3, 3A, or 3-A?), and deserve research and clarification during the intercensal decade. We further discuss the particular problem of small multiunit structures, whose division into subhousing units is not obvious, in Section 5–C.1.

A second problem that compromised the MAF in the 2000 census was housing unit duplication. Duplication is an ever-present problem in a resource like the MAF, which is an amalgam of various sources. In 2000, evidence of housing unit duplication prompted an unplanned, ad hoc process—effective but risky—to filter potential duplicates during the actual conduct of the census (Nash, 2000). Precise GPS coordinates may be useful in identifying some duplicates (e.g., structures at a street intersection that may be recorded on one street in one source and on the cross street in another source), but broader, more systemic sources of duplication should also be a focus of research and evaluation in preparation for the 2010 census. As we will discuss later, identification of MAF input sources that contributed duplicate addresses should provide vital evidence in remedying duplication problems.

5–B.2 Group Quarters

The logical counterpart to a call to reexamine the basic definition and interpretation of housing units is to do the same for group quarters. The group quarters population is people who live in such places as college dormitories, prisons, nursing homes, juvenile institutions, long-term care hospitals and schools, military quarters, group homes, shelters, and worker dormitories, and not in individual apartments or homes. In Census Bureau terminology, individual group quarters (e.g., a dormitory) may be nested within a broader construct, called a *special place* (e.g., a university).

In 2000, as in past censuses, enumeration of the group quarters population followed a separate operational plan from the rest of the census, using slightly different variants of the census form depending on the type of group quarters. However, National Research Council (2004:Sec. 4–F) suggests serious operational problems in group quarters enumeration, including the following (see also Jonas, 2002, 2003):

- *Failure to reconcile MAF development and group quarters listing processes:* The distinction between some group quarters (e.g., halfway homes and some assisted-living facilities) and regular housing units is not always clear, and there was an incomplete effort to produce—between the MAF and special place/group quarters roster—an accurate, nonoverlapping, and comprehensive address list. Moreover, the opportunity for local and tribal governments to review the independently generated list of special places was severely hampered by a late start. Group quarters thus became an important source of duplication in the census. For example, most recently, the Census Bureau acknowledged that a 2,673-resident dormitory at the University of North Carolina at Chapel Hill had been double-counted—a highly contentious finding since North Carolina narrowly edged out Utah (by 857 residents) for the 435th and final seat in the U.S. House of Representatives (Baird, 2003).

- *Geographic location codes mishandled:* The Bureau's Count Question Resolution program, beginning in 2001 and continuing through 2003, offered local authorities the opportunity to challenge their census population counts for geocoding and other errors; if successful, they could receive a letter certifying their revised count, although the program specified that corrected counts would not affect either reapportionment or redistricting. As the program proceeded, it became clear that a number of group quarters—including long-established prisons and college dormitories—had been coded to the wrong town or county.
- *Ineffective processing:* Instead of a bar code tracking system for individual group quarters residents, the Bureau's processing relied on a total count of questionnaires logged on a control sheet; in late processing in May 2000, the Bureau had to mount a special review after it was discovered that many suspected group quarters questionnaires were not properly associated with such a control sheet and thus could not be verified as group quarters enumerations.

Moreover, National Research Council (2004:297–301) concluded that the 2000 census long-form data for the group quarters population was poor, suffering from very high missing data rates and raising the legitimate question of whether it was appropriate to publish those data.

From the 2010 census planning perspective, developing a new approach to dealing with group quarters is important in several respects, in addition to our paramount interest in a complete and accurate address list. Among these is the prospect for the American Community Survey as a replacement for the census long form. Though the resultant data in 2000 may have been poor (National Research Council, 2004), the census long form is currently the only way of obtaining detailed characteristics information on the group quarters population. Accordingly, to truly replace the long form, the American Community Survey must include the group quarters population and improve the resulting data from that population (see Section 4–F.1). Special place/group quarters redefinition and reexamination must be part of the broader strategy to counteract duplication (Section 5–E). To the extent that the 2010 census uses the same postenumeration survey/dual-systems estimation approach to coverage evaluation as did the 2000 census, it will also be important to consider incorporating group quarters residents in the postenumeration survey. One complication of the 2000 coverage evaluation operations was that individuals—in particular, students—might move from university housing to private housing between Census Day and the time of nonresponse follow-up, thereby moving from outside to inside the survey universe. This complicated the estimation of coverage error and increased the variances of the resulting estimates. By including group quarters residents in the postenumeration survey in 2010, this complication could be avoided.

National Research Council (2004) urged a top-to-bottom reassessment of the Census Bureau's handling of group quarters; we endorse and restate here a version of their recommendation.

> ***Recommendation 5.4:*** **The Census Bureau's techniques for enumerating the population in special places and group quarters must be completely evaluated and redesigned for the 2010 census. This effort must include (but not be limited to):**
>
> - **clear definitions of group quarters;**

- **redesign of questionnaire and data content as appropriate, including a provision for handling data items that might best be provided by group quarters administrators rather than individual residents;**
- **collection of information, including additional addresses, that will be needed to facilitate unduplication of all census records;**
- **improvement of the address listing processes for group quarters, including coordination with the development of the Master Address File; and**
- **specification of enumeration and coverage evaluation plans for group quarters.**

5–B.3 Residence and Residence Rules

The current definition of residence rules is confusing both to field enumerators and to residents. Difficulties arise for people with multiple residences, including those with movement patterns that are primarily within a week, or those that move seasonally. Such movement patterns are typically true of retirees, those involved in joint custody arrangements, those with weekend homes in the country, students away at college during the school year, and people temporarily overseas. Clarification of residence rules in the census, and identification of better means of collecting information for later assessment, would aid census data collection.

5–B.4 Wording and Format of Race and Hispanic Origin Questions

A persistent and long-standing definitional concern arises from the Census Bureau's offering of "Some Other Race" among the possible answers to the question that elicits the racial composition of households. One obvious problem is the ambiguity of the category—it yields very little descriptive information about persons who provide this response. Another, more practical problem is that respondents who report "Some Other Race" must be reclassified for other uses such as obtaining denominators for vital statistics that do not employ this category. The use of "Some Other Race" is an especially acute problem for the Hispanic population, which is disproportionately represented in the "Some Other Race" category.

Studies from the 2000 census and the 2003 National Census Test (NCT) offer some insight into ways that it might be possible to reduce the numbers of persons reporting "Some Other Race." One clear finding is that the format of the 2000 census was a significant improvement over the format used in the 1990 census, because it reorganized the race and Hispanic origin questions and this reorganization resulted in a decline in both item nonresponse and the selection of "Some Other Race." Nonetheless, the use of "Some Other Race" was still widespread in the 2000 census, and so the Census Bureau has undertaken research to explore alternatives. The 2003 National Census Test examined a variety of different formats including questions that omitted "Some Other Race" as an option. While deleting this option necessarily lowers the numbers of persons selecting this category, it also increases item nonresponse in the race question. However, the item nonresponse does not outweigh the reduction in the selection of "Some Other Race," and as a result there is a net gain in the number of Hispanics using the standard OMB categories to report their race.

While it is tempting to conclude simply that deleting the "Some Other Race" option will resolve this issue, the National Census Test indicated that a sizable number of Hispanics—about 12 percent—will continue to report "Some Other Race" regardless of whether the option is offered. One clear implication of this result is that a sizable proportion of the Hispanic population does not envision itself in the same way that OMB chooses to partition the racial composition of the United States. One hypothesis that has not been well examined by the Census Bureau is that a number of Hispanics regard their national origin (e.g., "Mexican") or some other nomenclature (e.g., "Mestizo") as the most appropriate designation of their race and, absent such options, choose either to not respond to the race question or to identify themselves as "Some Other Race." Needless to say, embedding Hispanic origins within the race question may resolve some problems while it creates others. Logan (2003), for example, documents a significant amount of heterogeneity in the categories he calls Hispanic Whites, Hispanic Blacks, and "Some Other Race Hispanics." The inclusion of Hispanic origin items might obscure some of this heterogeneity. However, there is also a clear precedent for doing so as a variety of Asian national origins (e.g., Japanese, Chinese) are permissible responses in the race question.

The absence of a meaningful response for Hispanics, along with format changes in the race question, has led to yet another, less recognized though consequential, problem: the percentage of Hispanics identifying themselves as "American Indians" doubled between 1990 and 2000. One possible explanation for this anomaly is that the guidelines issued by OMB in 1997 stipulate that Central and South American Indians should be enumerated as American Indians. While this is anthropologically accurate, the problem it creates is that the governmental entities responsible for serving American Indians—as a result of treaty or other legal obligations—no longer have a straightforward count of the number of persons belonging to state and federally recognized tribes. While there has been a historic slippage between the numbers of American Indians enumerated by the census and those designated as members of federally recognized tribes by agencies such as the Bureau of Indian Affairs, the addition of Central and South American Indians makes these distinctions even more problematic.

In anticipation of the 2010 census, the Census Bureau needs to continue to study these issues carefully. In particular, it needs to devise a strategy for offering Hispanic respondents a meaningful option on the race question besides "Some Other Race." The deletion of this option without the substitution of a more meaningful response is a less than optimal solution. At the same time, the Bureau also needs to carefully examine the effects of format changes on other populations, such as American Indians, and pursue ways of providing data for American Indians that observes the "government-to-government" relationship pledged by presidential administrations since Richard M. Nixon.

As the Bureau conducts further research on race and Hispanic origin question wording and structure, those research efforts should also consider the impact of question design choices on the production of intercensal population estimates. This estimation task has become complicated by the revised question for eliciting racial background in the 2000 census. There are 126 unique combinations of race and Hispanic/non-Hispanic origins. For most areas of the country, and possibly for the nation as a whole, this is an intractable number of groups for which to provide estimates and could therefore serve to overwhelm the existing system for producing intercensal estimates. At present, the Census Bureau has managed this problem by reporting data for race and Hispanic origins for persons who report one race only ("race alone") and for the total population ("race alone" or in combination), thus making it possible to derive the multiracial population by subtraction. In the short term this is a reasonable stopgap measure and perhaps the only option currently available to the Bureau. However, in the long run, this is an unsatisfactory approach because the multiracial population is extremely heterogeneous and reporting data for this group in the aggregate masks many important interactions with respect to other characteristics, such as education and income. For example, a person who is black and American Indian may be very different in terms of a host of characteristics from a person who reports her or his race as white and Asian. At present, there is little reason beyond the convenience of available data for using the current approach. The Census Bureau needs to carefully scrutinize this practice in order to determine whether another set of categories might be more useful and hence more defensible.

The option for multiracial reporting also raises challenges of interpretation and analysis in the production of demographic analysis estimates, which have been a valuable benchmark in assessing the quality of census coverage. We discuss demographic analysis further in Section 7–B, and suggest that research on optimal ways to collect data on the multiracial population also consider implications for demographic analysis.

5–C HARD-TO-COUNT POPULATION GROUPS: EXTREMES OF URBANICITY

It is a basic truth that some population groups, some housing types, and some living structures are simply harder to count than others. Some population groups are not as amenable as others to mailout/mailback methods; some living structures are in areas where it is more difficult to generate and maintain an accurate address list. Examples of hard-to-count populations include the homeless, the high-density immigrant colonias along the U.S.-Mexico border, and residents of "gated" communities with restricted access.

We recognize the difficulty in crafting a strategy—*any* strategy—for enumerating these communities, and cannot suggest specific procedures for handling them. What we do strongly recommend, though, is that development of plans for these areas not be deferred to late in the process; strategies will be better informed by active research and evaluation, including reanalysis as appropriate of data from the 2000 census.

> ***Recommendation 5.5:*** **The Census Bureau's development of tailored enumeration methods for special populations—including irregular urban areas, colonias, gated communities, and rural areas—must begin early, and not be put off for development late in the census planning cycle.**

In the sections below, we offer further commentary on two such hard-to-count areas that we feel particularly strongly need early attention in research and planning; collectively, they represent the extremes of urbanicity.

5–C.1 Small Multiunit Structures and Immigrant Communities

Over the past three decades, immigration has become a more important component of population change. One-third of the population increase in the 1990s resulted from the increased presence of immigrants in the nation, not counting the added impact of higher fertility levels among immigrant women (Martin and Midgley, 2003). While a majority of the foreign born live in five states—California, Florida, Illinois, New York, and Texas—the pattern is changing, with more immigrants settling outside these states in the 1990s (Passel and Zimmerman, 2001). Immigrants are now found in many places in the nation, especially in large and small cities in neighborhoods where housing has been vacated by non-Hispanic whites. These immigrant settlements are frequently characterized by extensive family networks, higher-than-average fertility, and larger-than-average household sizes, all of which greatly complicate the census enumeration.

Many immigrant families also have English language proficiency problems, are fearful of government, and have occupancy characteristics that may violate local ordinances. All of these factors decrease the likelihood of their completing and returning a census questionnaire via mail, requiring high levels of nonresponse follow-up (NRFU). NRFU may be further compromised in these neighborhoods because their housing stock violates some of the basic tenets of the mailout/mailback method of data collection, the most important of which is the clear demarcation of housing units, especially in small multiunit structures. Small buildings that were once occupied by a single family are now home to multiple families in all kinds of configurations in many of the nation's cities, large and small. As the presence of immigrants and their children becomes more widespread, this problem will become more pronounced, threatening the most elementary assumption of a census enumeration—that it is possible to uniquely identify a housing unit for the purposes of mailing questionnaires and for conducting nonresponse follow-up.

In some neighborhoods, questionnaires can no longer be linked to housing units in any exact way, creating confusion about the delivery points for questionnaires and the completion of nonresponse follow-up. When a questionnaire fails to be returned by mail, the NRFU enumerator does not have a clear apartment designator in follow-up because such designators do not exist—mail is sorted by tenants of separate apartments out of a single mailbox, or, where multiple mailboxes exist, the use of apartment designators is inconsistent or nonexistent.

This situation leads to underenumeration and/or erroneous enumeration in the very places where effective counts are most needed for program planning and targeting. But although there have been calls for the Census Bureau to address this problem (see, e.g., National Research Council, 1999), the Bureau has failed to develop methods to deal with it, continuing instead to rely on a haphazard NRFU effort in immigrant communities that, in 2000, likely contributed to considerable undercount and erroneous enumeration.

The Bureau can no longer ignore this situation. The 100 percent block canvass and the actual census enumeration must employ new methods. No block canvass, regardless of the effort, will work if the rules regarding the listing of housing units do not take into account the occupancy and labeling problems that frequently characterize immigrant communities. The address listing operation assumes the existence of unit labels that are not present, so the very premise of the operation is faulty. The Bureau needs to create ways to label units and carry these labels into the enumeration. Mailout to these units may be impossible, so strategies need to be developed to take this into account. The Bureau needs to use the 2006 test as an opportunity to:

1. create labels for housing units in multiunit structures where no labels exist;
2. test methods for incorporating this labeling into a block canvass operation;
3. determine whether mailout can be conducted to these units; and
4. test an enumeration strategy that does not use the standard mailout/mailback method of data collection.

Alternate enumeration strategies might include urban update/leave, in which questionnaires are delivered to apartments by enumerators with a request that they be mailed back. This approach needs to include a component that labels the apartment, so that the questionnaire-apartment assignment is correct and so that follow-up can steer the enumerator to the correct location. Other options the Bureau should explore include more extensive use of face-to-face enumeration, in cooperation with local community leaders.

5–C.2 Rural Enumeration

The Census Bureau has historically been challenged by rural enumeration. Problems range from the absence of city-style address formats to physical barriers in remote, isolated places such as Alaska and the desert Southwest. Other challenges arise from a subset of rural respondents who live in these remote places because they are seeking to escape the intrusions of modern life, and especially the intrusions of the federal government. Finally, there is an often underappreciated diversity of places in rural areas—American Indian reservations, Hispanic colonias, and religious communities such as the Amish, to name only a few.

The Census Bureau should be mindful of two considerations in the enumeration of rural areas. First, it should avoid treating such areas as essentially homogeneous regions. Enumeration methods that work well on Indian reservations may not work well in rural Appalachia, and vice versa. Housing arrangements may vary from one rural area to another. Second, the partnerships formed for the 2000 census were instrumental in ensuring the cooperation of many rural communities. The partnership program and the many efforts made during the 2000 enumeration to "localize" the census and make it attuned to the interests of diverse communities should be carefully examined to build upon the successes of 2000 and to rectify any problems that may have arisen during the 2000 count.

5–D ALTERNATIVE RESPONSE MODES AND CONTACT STRATEGIES

Following up on limited experience in the 2000 census, the Census Bureau plans for alternative modes of response to the questionnaire to play a larger role in the 2010 census than they did in 2000. In particular, the response modes that have been proposed (in addition to mailback of the paper census form) are submission via the Internet and answers using an automated telephone system known as interactive voice response (IVR). As is the case with PCDs, the Census Bureau has suggested that increased usage of these response modes—both of which feature automated data capture, as data are collected in digital form—will achieve significant cost savings. As is also the case with PCDs, much remains to be demonstrated regarding the accuracy of these cost-saving assumptions; research is also needed to address the possible effects of alternative response modes on potential duplication in the census (see Section 5–E) and the potential for respondents to answer questions differently if the structure and wording of questions under the different response modes vary.

In addition to expanding the possible response modes, the Census Bureau has suggested that it plans to revise some of its respondent contact strategies. In 2000, advance letters were mailed before questionnaires were delivered and reminder postcards to send in the form were sent a few weeks after questionnaire mailout. The Census Bureau has expressed interest in sending a second questionnaire to nonresponding households, reviving an idea that had to be abandoned in the 2000 planning cycle. We discuss both the response mode and contact strategy proposals in the balance of this section.

5–D.1 Response Modes in 2000 and Early 2010 Testing

Internet data collection was conducted in the 2000 census, albeit in a limited and unpublicized manner. The Bureau's evaluation report on 2000 census Internet collection notes that there were 89,123 Census ID submissions on the Web site, of which 16.7 percent were failures (thus 0.07 percent of eligible households—63,053 out of 89,536,424 short-form households—successfully availed themselves of the opportunity to complete the form on the Internet) (Whitworth, 2002). This seems like a relatively high failure rate, although the report notes that "many, if not most, of the submission failures were associated with a Census ID representing a long form" (Whitworth, 2002:5). Given the Bureau's plans to expand the use of Internet reporting for the 2010 census, it is important to examine the data from the 2000 Internet responses, as well as from the 2003 National Census Test, to identify and correct problems such as those relating to entering the ID or other security or usability issues. We urge the Bureau to examine the data already in hand with a view to improving the design of the Internet response option.

The 2000 census also included in its evaluations and experiments program a Response Mode and Incentive Experiment (RMIE), testing whether respondents would be more likely to submit a census form via the Internet or IVR if offered an incentive (specifically, a telephone calling card). Among those assigned to Internet (i.e., encouraged to complete the form on the Web), only 3.9 percent did so when given an incentive to do so, and 3.4 percent did so with no incentive. Extrapolating these numbers to the entire set of eligible households in 2000, providing an incentive to use the Internet option would have resulted in just over 3 million returns by this mode. The summary of the RMIE work suggests a potential saving of between $1 million (assuming a 3 percent Internet response) and $6 million (assuming a 15 percent response) in postage costs (Caspar, 2003). The Bureau has argued that savings in paper, printing, data capture, and warehousing costs resulting from converting many mail responders to alternative electronic response modes such as the Internet would help offset the costs of acquiring PCDs. Given the above numbers, we do not see large potential savings from alternative response modes and we urge the Bureau to develop realistic cost models for such approaches.

Caspar (2003), summarizing the various RMIE reports, offers other insights into the potential effectiveness of alternative response modes in 2010:

- The calling card incentive moved some people to use the alternative mode, but did not increase overall response as these are people who would respond by mail anyway.
- The impact of the calling card incentive may not justify its cost.
- Among respondents to the Internet Usage Survey who were aware of the Internet option, 35 percent reported that they believed the paper form would be easier to complete. While Internet completions may be beneficial for the Census Bureau, the argument needs to be made for its benefits to the respondents before large numbers of them are likely to switch to Internet completion.
- IVR does not look promising: "Without significant improvements in the voice-user interface, the IVR technology is probably not a viable option for Census 2010."

The results from the alternative mode part of the 2003 National Census Test support the view that alternative modes are unlikely to account for a large proportion of census responses or to decrease mail nonresponse in 2010: "Offering a choice of alternative modes did NOT increase or decrease the cooperation rates. Instead it simply shifted response to the alternative modes. However, this shift was relatively small. . . . Pushing respondents to respond by electronic modes [IVR or Internet, by not providing an initial paper questionnaire] was found to decrease overall response" (Treat et al., 2003:8). The response mode component of the 2003 test is also discussed in Box 9.1 in Chapter 9.

5–D.2 Response Mode Effects

It is well known in survey research that respondent answers may differ due to variations in the precise wording, format, and structure of the questions, and may differ based on the mode in which the survey is rendered (e.g., self-response versus interviewer-administered). Reporting the results of the 2003 National Census Test, Treat et al. (2003:9) recommend that the Census Bureau "develop content suitable for each mode; we need to first develop an ideal instrument for each mode, then test against other modes for differences." As the Bureau report notes further, redesigns of the questionnaires used under different response modes should be sensitive to the possibility of mode effects on respondent answers; specifically, the report recommended "research[ing] the design of the instruments so as not to compromise data quality, while maximizing the advantages of each mode" (Treat et al., 2003:9).

Assuming a short-form-only decennial census, concern over mode effects is eased somewhat due to the nature of the questions; several of the basic data items such as gender and housing tenure are not likely to be hurt by nuance in wording and format. The race and Hispanic questions, however, are a key area of possible concern for response by mode. This is particularly true given the multiple-response form of the race question and the demonstrated sensitivity of the Hispanic origin question to the number of groups mentioned as examples, as suggested by the 2003 National Census Test (Martin et al., 2003; see also Box 9.1). The ACS, consisting of the current long-form-sample data items, is more sensitive to mode effect concerns, though the extent to which alternative response modes may be added to the

ACS plan is unknown. Of greater concern with respect to the ACS are potential differences in response that may arise from different question structuring between the ACS instrument and the current census long form.

5–D.3 Replacement Questionnaires

Prior to the 2000 census, the Census Bureau's initial plans to send a replacement questionnaire to nonresponding households had to be abandoned after it was determined that the operation could not be completed in a timely manner. While the address list for targeted nonrespondents could be developed quickly, the Bureau learned from contractors that the actual printing, addressing, and mailout of questionnaires would take several weeks, delaying any nonresponse follow-up effort by an unacceptable amount.

The results of the 2003 National Census Test were consistent with previous results in the survey literature, showing that targeted replacement questionnaires had a significant effect on cooperation rates—a 10.3 percentage point increase at the national level. The panel is convinced that the potential effect of replacement questionnaires on mail response rates has been well demonstrated and that implementation of this contact strategy in 2010 would be beneficial. What is needed now is a specific operational plan in order to actually deliver the replacement questionnaires.

Treat et al. (2003:11) comment that, "in the Census, the largest obstacle for a targeted replacement questionnaire to nonresponding households is how to operationalize it." It is certainly not the only obstacle; both the replacement questionnaires and the greater use of alternative response modes increase the potential risk of duplicate enumerations, and so development of strategies for unduplication becomes increasingly important (see Section 5–E). Furthermore, as Treat et al. (2003:11) note, rcsearch also remains to be done on the optimal time lag after the initial questionnaire mailout to compile the list of nonrespondents and send replacement questionnaires.

> ***Recommendation 5.6:*** **The Census Bureau must quickly determine ways to implement a second questionnaire mailing to nonresponding households in the 2010 census, in order to improve mail response rates. Such determination should be done in a cost-effective manner that minimizes duplicate enumerations, but must be made early enough to avoid the late problems that precluded such a mailing in the 2000 census.**

In the panel's assessment, research consideration of possible effects of response mode and questionnaire design on respondent answers is certainly warranted and should be pursued. That said, a more pressing concern is development of plans for dealing with census duplication and nonresponse, as we describe in the next section.

5–E DATA-PROCESSING METHODOLOGIES: UNDUPLICATION AND IMPUTATION

Two basic data-processing stages became very prominent in the 2000 census experience, and are likely to remain so in 2010. *Unduplication* (referring here to person records) became a major focus of the follow-up research informing the various decisions on possible statistical adjustment of the 2000 census totals. Specifically, advances in unduplication were made possible by a reasonably simple innovation—name and date of birth were captured for the first time in the 2000 census, as a byproduct of the use of optical character recognition technology. Based on work by Fay (2001), the Bureau staff continue to use and enhance the capacity to search the nation for individuals matching on name and date of birth. Especially for very common names, some of these matches are false, but weighting procedures have been developed to account for false matches. There is the real possibility of using some variant of this procedure to substantially reduce the frequency of duplicates in the 2010 census and coverage measurement program.

Likewise, *imputation* for nonresponse emerged as a major focus in the wake of the 2000 census. The Census Bureau's basic methodology for imputing missing data items—so-called "hot-deck" imputation—has been in use for some 30 years. And, though it has certain key advantages—among them that it can be performed in one pass through the data set—it is a methodological area ripe for new research and approaches. Imputation gained considerable attention in the 2000 census when the state of Utah questioned its use in the second of the state's major legal challenges against the census counts, arguing that imputation constituted statistical sampling (which is prohibited from use in generating apportionment totals). The U.S. Supreme Court rejected the argument, ruling that "imputation differs from sampling in respect to the nature of the enterprise, the methodology used, and the immediate objective sought" and that use of imputation is not inconsistent with the "actual enumeration" clause of the U.S. Constitution (*Utah v. Evans*, 536 U.S. 452, 2002). Though imputation methods withstood legal scrutiny in this instance, their use and the potential implications they bring will likely be the subject of continued debate.

National Research Council (2004) offers three recommendations related, generally, to the Census Bureau's plans for unduplication and imputation in the 2010 census. We endorse and restate them here.

> ***Recommendation 5.7:*** **The Census Bureau must develop comprehensive plans for unduplication in the 2010 census, in terms of both housing units and person records. Housing unit unduplication research and efforts should be conducted consistent with objectives outlined in the panel's recommendations related to the Master Address File. Person-level unduplication efforts should focus on improvements to the methodology developed for the 2000 Accuracy and Coverage Evaluation Program, including national-level matching of records by person name. It is essential that changes in unduplication methodology be tested and evaluated using extant data from the 2000 census and that unduplication methods be factored into the 2006 proof-of-concept test and 2008 dress rehearsal.**

Recommendation 5.8: **The Census Bureau must pursue research on the trade-off in costs and accuracy between field (enumerator) work and imputation routines for missing data. Such research should be included in the 2006 proof-of-concept test, and census imputation routines should be evaluated and redefined prior to the 2008 dress rehearsal. As appropriate, the American Community Survey research effort should also address the trade-off between imputation and field work.**

Recommendation 5.9: **The Census Bureau should conduct research into the effects of imputation on the distributions of characteristics, and routines for imputation of specific data items should be completely evaluated and revised as appropriate for use in the American Community Survey.**

CHAPTER 6

Technical Infrastructure and Business Process

MASSIVE AMOUNTS OF INFORMATION—from the answers to every question on every returned questionnaire to the personnel and payroll records for hundreds of thousands of temporary employees—must be managed and processed in order to conduct a successful decennial census. To process all this information, the census relies on a complex technical architecture—the collection of people, computer hardware and software, and telecommunication networks that supports the complete workings of the census. Included in this technical infrastructure are subsystems to track personnel hires and fires, monitor caseload and make enumerator assignments, capture and synthesize data, generate maps, and myriad other functions, which must function not only at Census Bureau headquarters but also at regional offices, data collection centers, and hundreds of temporary local census offices.

The 2000 census relied on several major systems (U.S. Census Bureau, 2000; Titan Corporation, 2003), including the following:

- *Geographic Support System (GSS):* a facility for deriving extracts from MAF/TIGER as necessary and printing enumerator maps;

- *Pre-Appointment Management System/Automated Decennial Administrative Management System (PAMS/ADAMS):* a system to support the hiring, processing, and payment of temporary employees, as well as administrative data archiving;
- *Operations Control System (OCS 2000):* a caseload management system to define and track enumerator assignments, as well as to monitor duplicate and missing addresses;
- *Data Capture System (DCS 2000):* a system for the check-in and scanning of completed questionnaires;
- *Telephone Questionnaire Assistance/Coverage Edit Follow-Up (TQA/CEFU):* a program to provide support for respondents requiring assistance or additional forms, as well as follow-up data collection from respondents by phone;
- *Internet Data Collection/Internet Questionnaire Assistance (IDC/IQA):* a system for the support of limited-scale Internet response to short-form questionnaires;
- *Accuracy and Coverage Evaluation (ACE):* a program to provide support for a follow-up survey to assess possible undercount (including maintenance of laptop computers used by enumerators and the Matching and Review Coding System [MaRCS] used in matching the survey responses to census returns);
- *Management Information System (MIS 2000):* a system for senior management planning and information tracking, including schedule and budget planning and tracking;
- *Headquarters (HQ) Processing:* the analysis and processing of final data, including production of reapportionment and redistricting population counts, as well as other data products; and
- *Data Access and Dissemination System (DADS):* a system for dissemination of census data to the public, most notably through the American FactFinder Web site (http://factfinder.census.gov).

In the end, the information systems of the 2000 census achieved the desired results. "Operationally, most agree that this decennial census was a success—participation was higher than anticipated ... and operations concluded on time," notes an assessment prepared by the U.S. Department of Commerce, Office of Inspector General (2002:iii). However, the assessment continues, the means by which it was achieved—including the patchwork of information systems—led to other descriptions: "costly, complex, and high risk."[1]

The technical infrastructure of the 2000 census was generated without reference to an overall blueprint; individual systems were pieced and linked together, often having been developed quickly and without full opportunity for testing. Though not as well publicized as the Census Bureau's major proposed initiatives for the 2010 census, the Bureau has taken steps toward a more rigorous development process for the 2010 census technical infrastructure. Specifically, efforts are under way to model the logical infrastructure of the census—the complete mapping of information flows through the entire decennial census. Properly executed, logical infrastructure models allow for alternative organizational structures and assumptions to be tested in the abstract. Alternative models can be compared before deciding on a model; that finished model then serves as blueprint, specification, and template for constructing the physical (hardware/software) technical systems. Full use of logical architecture modeling has the potential to greatly reduce risk in system development and ensure that the various information subsystems of the census communicate effectively with each other.

In this chapter, we examine this modeling effort as well as the Bureau's broader effort to develop its technical infrastruc-

---

[1]The "high risk" nature of system operations is illustrated in the following example. In late 1999, the Commerce Department's Office of Inspector General reviewed one of the constituent information systems of the 2000 census—the PAMS/ADAMS system to track personnel hiring and payroll. Based on interactions with the Census Bureau, the report concluded that the Census Bureau "did not follow a well-managed software development system" in creating PAMS/ADAMS, but the Bureau was confident that the system would be able to support decennial census operations given "extensive operational use" of the system since the 1998 dress rehearsal. By January 2000, further review led the Bureau to conclude that the PAMS/ADAMS might *not* be fully capable to support decennial needs and undertook "extensive software modifications" less than 3 months before Census Day (U.S. Department of Commerce, Office of Inspector General, 2000:i–ii).

ture. Section 6–A describes the basic concepts of an architectural model and discusses the Census Bureau's initial implementation; our assessment of the modeling effort is given in Section 6–B. In Section 6–C, we address a major, specific piece of the broader architecture for 2010: namely, the revised database structure for the Master Address File (MAF) and TIGER system. We close the chapter in Section 6–D by outlining major challenges faced by the Bureau in managing and finalizing the technical infrastructure of the 2010 census.

6–A TOWARD A "BUSINESS PROCESS" OF THE DECENNIAL CENSUS

Past experience with reengineering and upgrading information technology operations within corporations and government agencies suggests that the most prudent and productive approach is to proceed in well-thought-out stages or steps:

- *Define a "logical architecture" or "business process" model.* A first step is to articulate the set of activities and functions currently performed by the organization and the informational dependencies among them. This model of activities and functions is called a logical architecture. It may also be called a business process model because it defines the ways in which operations are carried out to accomplish the intended objectives of an organization. In the census context, the current business process would be the information flows and tasks associated with the 2000 census. We will explain the nature of logical architecture or business process models in greater detail in the following section.
- *Reengineer the logical architecture.* The completed logical architecture may be viewed as an "as-was" model; again, in this case, the as-was model would describe the activities of the 2000 census. Using the as-was model as a base, the next step is to produce one or more "to-be" models—that is, to identify new assumptions and objectives and to adjust the as-was logical architecture model as necessary to find the optimal way to structure functions under the new demands.

Different to-be models can then be compared against each other in order to reach a final architecture model.

- *Construct the physical technical infrastructure using the reengineered logical architecture as a guide.* The finished logical architecture/business process model is then used as the template and specification for a new physical technical infrastructure—the actual network of hardware and software systems assembled to carry out the organization's work.

Any other approach—such as failing to map business functions in terms of overall objectives or rushing to make decisions on technical infrastructure too early—serves only to allow the organization to make more mistakes, albeit (probably) faster than before.

The Census Bureau has begun the task of reengineering the decennial census infrastructure in this manner because it fits into the objective of early planning and testing envisioned as part of its broad strategy for the 2010 census and because it brings the Bureau and the Department of Commerce into fuller compliance with the Information Technology Management Reform Act of 1996 (also known as the Clinger-Cohen Act).[2] This act called for federal agencies to reexamine their information technology (IT) structures, requiring greater attention to how IT furthers the agency's goals and to modeling current and modernized IT structures as a business process. The Chief Information Officers (CIO) Council, created by executive order, subsequently developed the Federal Enterprise Architecture Framework (FEAF), a set of minimum standards for description of IT programs and modernizations.

6–A.1 Baseline: Logical Architecture of the 2000 Census

The Census Bureau contracted with the Centech Group, an IT company based in Arlington, Virginia, to develop its baseline for infrastructure reengineering: namely, a business process model

[2]The Information Technology Management Reform Act of 1996 is part of Public Law 104-106. Among other provisions, the act also encourages the use of commercial off-the-shelf (COTS) products instead of or in conjunction with software systems built within government agencies.

of the operational flows underlying the 2000 census. Lockheed Martin was subsequently brought in as a subcontractor. The result of this first stage of work is a map of the logical architecture of the 2000 census, and it is summarized in a report by the contractor (Centech Group, Inc., 2002a). A more detailed companion volume examines each logical segment of the model in greater detail (Centech Group, Inc., 2002b). The model developed in this contract does not cover every decennial census operation but concentrates on what the Census Bureau identified as major business process areas.

The logical architecture models developed by the Census Bureau under this contract adhere to the Integration Definition for Function Modeling (IDEF0) language, a method that has been adopted as a federal standard for representing organizational functions and flows.[3] IDEF0 models use simple graphical structures to organize information. Functions (activities) of an enterprise are rendered as boxes, which are connected by arrows representing information constraints. For large enterprise models, a high-level diagram is typically produced as a guide or road map for the analyst; smaller pieces are then indexed based on this high-level map and are available in full detail on separate pages.

A logical architecture model is a blueprint of the workflow of a particular enterprise. It describes the nature of information that must be passed from point to point at various phases of the operation and, in doing so, highlights information interfaces—points of connection both within the system and with external entities. The model thus defines the baseline capability that must be present when a physical technical infrastructure is constructed. It may also convey a rough sense of where, geographically or organizationally, groups of activities should be clustered.

To better understand what a logical architecture model of the decennial census *is*, it is also important to be clear about what it is *not*. The main purpose of an IDEF0-based logical architecture model is to emphasize process and function. To that end, the model effectively disregards two variables that are

---

[3]Specifically, IDEF0 was released as a standard in 1993 in Federal Information Processing Standards (FIPS) Publication 183.

of some natural concern. First, it does not assign completion times to any function or process. Rather, it describes forward information flow through a business process without delineating a timeline or schedule of the process. Individual segments of the model may be completely distinct in terms of their execution time or may overlap extensively. Second, IDEF0 models are not based on existing organizational boundaries; logical segments are partitioned strictly based on function and purpose, without respect to internal work divisions that may already exist within an enterprise.

An important question in building IDEF0 models is the level of detail required in the diagrams in order to facilitate effective process reengineering. FIPS Publication 183, which defines IDEF0 structures, suggests that each parent box (function) be decomposed until it can be expressed in 3 to 6 child boxes (Part B.2.1.4). The arrows representing information constraints should be expressed in the same level of detail as the boxes (Part B.2.2.2); that is, a rule of thumb is that activities are not adequately decomposed if boxes have more than 6 arrows on any side.

Finally, since the concepts may be confused, it is important to emphasize that a logical architecture is not equivalent to a physical computing or technical architecture. Properly executed, a logical architecture does not define the specific computing platform or database structure to be used, and it certainly does not presume to dictate the specific variables or records to be saved in particular databases. However, the logical architecture can provide a template for the physical trappings; the diagrammed flows and constraints of the model give shape to and provide baseline specifications for the types of activity that physical systems must be able to perform. Moreover, although a logical architecture documents work, it should be invariant to specific operational decisions—whether certain data are input at one computer or at twenty or, in the context of the census, whether operations take place in 500 local census offices or 600.

After defining operational flows, the Census Bureau began to render diagrams and logical flows captured in the logical architecture model for the 2000 census using System Architect, a software package developed by Popkin Software, Inc. This work was done in support of a limited pilot "reengineering exercise."

6–A.2 Reengineering Exercise

Between August and October 2002, Census Bureau staff performed a logical architecture reengineering exercise, again contracting with the Centech Group, which issued the final results in a report (Centech Group, Inc., 2002c). To keep the exercise manageable, given the Bureau's newness to the process, reengineering activities were narrowed in scope to focus on the census process steps from data collection through data processing. Candidate areas for retooling were proposed and considered for inclusion in the exercise, which ultimately concentrated on adapting the as-was model of the 2000 census to reflect three potential areas of change:

- *Control of follow-up procedures:* make nonresponse follow-up assignments dynamically, based on regular updates of response status for all housing units during census conduct and on the progress of individual enumerators;
- *Centralized data capture and formatting for all response modes:* ensure that data provided to headquarters are in uniform format regardless of response type (mail, telephone, Internet); and
- *Redistribution of "undeliverable as addressed" questionnaires:* adapt sorting and screening processes to streamline handling of questionnaires returned by the U.S. Postal Service, for easier identification of vacant housing units.

Architecturally, adaptation of the as-was 2000 census model to incorporate these operations included many changes in follow-up information processing as well as the addition of data centers[4] to perform processing and formatting tasks.

As part of the exercise, Census Bureau staff developed a list of sixteen principles to guide the logical architecture as the three

[4]Here, "data center" refers to a designated point to handle sorting and reformatting tasks. Use of the term should not be confused with the Census Bureau's state data centers, which are part of the apparatus for data and analysis outreach to users.

selected changes were incorporated in a to-be design. As the contractor report notes, individual architectural principles may, by design, oppose each other—"optimization for one principle may cause non-compliance with another principle" (Centech Group, Inc., 2002c). The hope is to find alternative architectural flows that best balance the demands of the entire set of principles.

In the Bureau's exercise, two of the architectural principles are "consider the needs of the respondent" and "facilitate counting everyone once, only once, and in the right place." These principles can be weighed against each other by the degree to which they contribute to overall goals of the enumeration. They can also be used to evaluate competing "to-be" logical architecture models. For instance, a higher number of response modes available to respondents under one plan might be considered evidence in its favor with respect to the "consider the needs of the respondent" principle, but not in its favor with respect to the "once, only once, and in the right place" principle due to the potential for duplication. In the reengineering exercise, Census Bureau staff identified a number of such measures (quantitative and qualitative), which serve as evaluation criteria to compare the baseline as-was model (the 2000 census structure) with the proposed initiatives for the 2010 census.

6–A.3 After the Pilot: Steps Toward an Architecture

Work on the pilot reengineering exercise ended in October 2002, and in January 2003 Census Bureau staff began work on other architectural products. Initial work on an activity model for the 2010 census was completed in October 2003 (U.S. Census Bureau, 2003b).

6–B ASSESSMENT

The panel enthusiastically endorses and supports the Census Bureau's work on its pilot logical architecture project and strongly urges its continuation.

Completion of a logical architecture model for the 2000 decennial census and of a redesigned model for the 2010 census would be major accomplishments and deserve recognition for their potential utility. As the contractor's report notes, the Census Bureau has traditionally put "little emphasis on assessment of the entire 'end-to-end' decennial census process" (Centech Group, Inc., 2002a:vii). Hence, the Bureau's efforts in working toward that complete model are indeed very encouraging. As we noted in our second interim report (National Research Council, 2003a), the Bureau's selection of modeling products and paradigms have thus far been quite sound.

6–B.1 The Need for Institutional Commitment

The Census Bureau's emerging plans for the 2010 census are laden with new initiatives and new technologies: a parallel data process in the ACS; more extensive ties to an updated MAF/TIGER system; data capture and transmissions from PCDs; Internet transactions; use of administrative records systems; and in-time collection and archival of information for immediate use in quality control and quality assurance. Each of these activities will require care when incorporated into a logical architecture for the 2010 census.

Constructing an extensively reconfigured logical architecture—and, more importantly, using the resulting model as a template for building the actual physical infrastructure for the 2010 census—is an arduous task. And though the effort of using a completely realized logical architecture to build the physical technical architecture will ultimately reduce operational risk in census conduct, the architecture-building process is not without risks of its own. In terms of general recommendations as the Census Bureau continues with its architecture work, the panel's suggestions are generally consistent with those of an earlier National Research Council panel on which members of the current panel also served. The earlier panel was charged to advise the Internal Revenue Service on the modernization of its internal systems (National Research Council, 1996), a task similar in certain respects to reconfiguration of the decennial census. Accordingly, our lead recommendations are similar. First, successful reengineering efforts typically require active "champions" at the highest management levels, and the Bureau must seek champions for its architecture construction process.

Second, in order to conduct a successful reengineering process, the Census Bureau will need to bolster its technical expertise in enterprise modeling.

6–B.2 Management "Champions"

The major technological enhancements envisioned under the Census Bureau's proposed plan for the 2010 census are distinctive not only for their range but also for the manner in which they cut across long-standing organizational divisions within the Bureau. For example, PCDs with GPS receivers are a field data collection tool, and therefore many requirements for the devices will have to be driven by field personnel needs; however, they are of limited use if the positional accuracy of TIGER is not improved. Additionally, computer-assisted questionnaires contained on the devices would benefit from cognitive and usability testing.

The approach of enterprise or logical architecture modeling is to concentrate on function and information flow rather than on preexisting work conditions, though indeed the finished result of modeling may suggest more efficient ways to structure operational workload. However, experience in carrying out similar infrastructure remodelings suggests that it will be vitally important to have strong support at the highest levels of management at the Bureau—in effect, to have influential "champions" of architecture reengineering. These people can effectively convey the importance of the task and encourage all divisions to "buy in" to modeling activities, and can then coordinate and integrate the emerging system.

> ***Recommendation 6.1:*** **In order to achieve the full benefit of architecture modeling, the highest management levels of the Census Bureau should commit to the design and testing of a redesigned logical architecture, so that the most promising model can facilitate the implementation of an efficient technical infrastructure for the 2010 census.**

6–B.3 Establishing a System Architect

The development of an adequate business process model for the 2010 census will require a serious effort that must be well staffed and well supported. Although the support and commitment of top-level management are necessary, the panel believes that authority for coordinating and developing that model should be vested in one person—a system architect for the 2010 decennial census. We recommend that such a position be created as soon as possible and that a well-qualified candidate be hired to fill the job.

The system architect should be supported by a full-time staff of reasonable size in order to ensure the expertise necessary for a modeling methodology that is new to the Census Bureau. The system architect and related staff have a primary role as information gatherers, tapping the expertise of other Bureau staff to build and revise architecture models. But another important role is outreach, in a sense—helping to build commitment to architectural principles by informing other parts of the Census Bureau of modeling results and demonstrating their usefulness.

As we will discuss in Section 6–C, a system architect has been appointed to oversee the redesign of the MAF/TIGER database redesign (Objective Two of the MAF/TIGER Enhancements Program). In our assessment, this is a positive development; the database redesign is a critical 2010 census activity, and strong coordination is helpful. We urge that the decennial census architecture and MAF/TIGER database redesign teams not work in isolation from each other; rather, their activities should be coordinated through regular interaction between the appointed system architects. The development of PCDs and other field systems is also a sufficiently major piece of the broader 2010 census architecture that we believe appointment of a subsystem architect could be beneficial.

> ***Recommendation 6.2:*** **To ensure the successful integration of new technologies and techniques in the census process, the Census Bureau should create and staff the position of system architect for the decennial census. The selected candidate should have exper-**

tise in modeling business processes, designing large-scale systems, and conducting reengineering activities. The system architect must be given the authority to work with and coordinate efforts among the organizational divisions within the Census Bureau and should serve as a champion of the importance of architecture reengineering at the highest levels of management within the Bureau.

The Census Bureau should also consider designating a subsystem architect for portable computing devices and related field systems, as has already been done for the MAF/TIGER redesign. The efforts of the MAF/TIGER redesign and PCD subsystem architects should be coordinated in partnership with the system architect for the decennial census.

6–B.4 Cautionary Note: Breadth and Difficulty of Task

We wish to make clear our view that it is both important and appropriate that the Census Bureau is pursuing enterprise architecture modeling of the decennial census. Proper execution of this modeling will facilitate the testing and evaluation of alternative system structures in the abstract, adding rigor to the development of census hardware and software support systems and reducing overall operational risk. But, to underscore the recommendations made in the previous sections, we also wish to make it clear that the difficulty of the task should not be underestimated, nor should the importance of championship and commitment to the modeling activity at all levels of the Bureau.

We have reviewed U.S. Census Bureau (2003b), the Bureau's "Business Architecture 1.0," as well as the reports from the earlier pilot logical architecture and reengineering studies. We have heard initial plans for the MAF/TIGER database modernization (Section 6–C) and have seen basic operational workflows for PCDs as they will be implemented for nonresponse follow-up in the 2004 census test (Section 5–A.1). That said, given the experience of some panel members in working on architectural reengineering of major systems, our impression—and it is admittedly only an impression—is that the components we have seen make up a rather small share (perhaps 20 percent or less) of the real architecture required to support the 2010 decennial census. It is also possible that—given the inherently limited nature of pilot activities to date—the products we have seen may reveal only some 20 percent of that 20 percent.

The 2010 decennial census, as a whole, must be viewed as a complex system; integration of that system has often been stated as a primary goal of the reengineered census. It must be recognized that all information systems employed during the decennial census are parts of the overall technical infrastructure that is necessary to support the census. But technical infrastructure "integration" cannot mean just providing a means for moving information back and forth among information subsystems—all systems are "integrated" by that limited definition. Rather, effective integration involves careful analysis of the distribution of functionality among subsystems, their informational interdependencies, and ultimately their geographical replication and distribution; it means careful examination for efficiencies and reduction of redundancies in task.

Our reading of U.S. Census Bureau (2003b) suggests that it represents a good start to building architectural models but one that can be improved with experience. In particular, the diagrams in the document are good at modeling the fine-level detail of various activities but are less good at giving a sense of context and placement within the system as a whole. They are not rendered at the level of detail that is appropriate for effective process reengineering (see guidelines in Section 6–A.1). For instance, a single diagram covering "Infrastructure" (basically, the actual building of the Bureau's information technology systems) shows 8 main activities; 81 information products are shown to be used or produced by these activities, and 24 supporting tools or systems are identified in the diagram. Of the 8 activities, only one—"Perform Logistics Support"—is decomposed in a finer-level diagram, while other activities such as "Manage Public Communication Program" and (particularly) "Manage Temporary Workforce" should probably be decomposed by another level or two before real reengineering of this model segment can fruitfully proceed (U.S. Census Bureau, 2003b:Tab 18, p. 7, Diagram A2). As it stands now, the diagram is too cluttered to convey high-level flows but not decomposed enough to uniquely identify all information flows. [Other similar examples can be found in the draft architectural documents, including U.S. Census Bureau (2003b:Tab 18, p. 9, Diagram A3), which depicts 8 activities, 163 information products, and 52 supporting tools and systems—far too busy as a high-level summary.] As the Census Bureau becomes more familiar with enterprise architecture modeling capabilities, we encourage it to consider a slightly more "top-down" approach in its modeling, revisiting and revising the high-level connections with activities and working down to the finer-activity details.

6–C THE ARCHITECTURE OF CRUCIAL SUBSYSTEMS: THE TIGER REDESIGN

Before we list additional comments and recommendations on the emerging technical infrastructure plans, we think it is useful to devote some attention to a specific, major piece of that larger puzzle.

Objective Two of the MAF/TIGER Enhancements Program is to convert the current database structure underlying the Master Address File and the TIGER geographic database to a modern processing environment. As we discussed in Section 3–A.2, the TIGER database was a considerable technological achievement when it was developed in the mid-1980s in support of the 1990 census. It was created by the Census Bureau using homegrown structures, in large part because commercial database applications available at the time were not well suited to managing the required topological integrity—the complex interrelations of various points, lines, and polygons that make up a national map. In the decades since, database software has made considerable advances while the TIGER database structure has remained largely the same. As a result, TIGER now suffers from archaic restrictions on file access and from difficulties in training staff to use the custom software. As an added complication, the Master Address File and the TIGER database have previously been maintained as separate structures, connected when necessary by geocoding (literally, referencing to find where address entries are located relative to TIGER-defined lines and polygons). The major aims in modernizing the MAF/TIGER database environment are to make the databases easier to maintain and use, as well as to establish a more rigorous link between the two by housing them in the same data structure.

During the panel's early interactions with the Census Bureau regarding the MAF/TIGER Enhancements Program, the difficulty of this Objective Two modernization seemed to be consistently underestimated. In those early discussions, the conversion was characterized as a fairly easy step: a new database structure would be identified and new support software would be written (and tested and certified error free). Work on the TIGER database could then be suspended for a period of a few days, information ported over to the new structure, and the task would be done. All experience with such upgrades suggests that such a rosy scenario is misguidedly optimistic. In our more recent discussions, the Census Bureau has, we are pleased to note, moved away from this earlier position and has made progress in defining and articulating the conversion task.

The Census Bureau's current plans are to complete the database conversion in fiscal 2006. To that end, fiscal 2004 will be particularly critical, with decisions slated to be made on the commercial off-the-shelf software packages to be adopted for use in the project and on requirements and specifications for both the hardware and software scheduled to be completed. Under the Census Bureau's current plan, fiscal 2005 would involve software development and testing and the installation and testing of necessary computer hardware; fiscal 2006 would involve continued testing and, ultimately, migration of the data.

It is important to note that Objective One—the realignment of TIGER features—is not contingent on the completion of Objective Two. The Census Bureau and its Objective One contractor are intending to maintain updated and realigned TIGER files in the old TIGER database format until Objective Two is complete and the new structure is ready. To our knowledge, there is no expectation that—should the Objective Two database conversion be completed on schedule in 2006—the Objective One contractor would be required to switch formats and begin providing realigned files in the new TIGER format.

We articulated some principal advantages of the TIGER

database conversion in Section 3–B.2 and throughout Chapter 3. Among these are easier potential data interface with state and local governments, given that the current native TIGER database structure is inconsistent with modern geographic information systems (GIS) software. Also, a modern system built on commercial database software makes it easier to recruit and train employees, rather than requiring extensive retraining in an old and site-specific software environment.

However, the project entails risks in several key respects. MAF and TIGER are both so central to census operations—given their use in creating maps, extracting census and survey address frames, and geocoding—that severe risk and cost could be incurred if the conversion is delayed or cannot be completed in a timely fashion. Failure to adequately test the new MAF/TIGER hardware and software—or the lack of adequate time to perform such testing—could easily lead to serious bugs and errors that may only be detected after the conversion is complete and therefore could be very costly. Moreover, even if the Bureau's timetable for the modernization holds true and the project is completed in fiscal 2006, that will be too late for a fully implemented modern MAF/TIGER system to be incorporated into the 2006 census test. The Census Bureau has already confirmed to one of its oversight authorities that a completed MAF/TIGER redesign and implementation cannot be tested in 2006 (U.S. Department of Commerce, Office of Inspector General, 2003).

Another potentially serious risk lies in the ability of selected commercial off-the-shelf software to meet requirements. The Census Bureau has already indicated that it has chosen to implement the core of the new MAF/TIGER system using Oracle Spatial database tools. Our understanding is that Oracle Spatial is a relatively new addition to Oracle's database products. In addition, implementation of the new MAF/TIGER will depend heavily on an Oracle add-on product, Topology Manager, which we understand is still in beta testing. While there is no reason at this time to suspect that the Oracle tools will be problematic, given their newness, the risk must certainly be acknowledged. The importance of maintaining the topological integrity of TIGER cannot be overstated.

As is the case with the balance of the MAF/TIGER Enhancements Program, the panel supports the goal of Objective Two. Modernization of the technical underpinnings of the MAF and TIGER databases is essential to the continued usability of these critical data resources, as well as to facilitate more seamless interactions between the Census Bureau and state, local, and tribal government partners. In terms of content, echoing comments in Chapters 3 and 8, we strongly recommend that the Census Bureau carefully consider the data items it includes in the new MAF/TIGER database system rather than simply porting existing data into a new shell. In particular, greater attention should be paid to storage of metadata and changelog information in order to facilitate quick and effective evaluation—so that, for instance, it is possible to reconstruct rather than approximate the history of a particular address as it appears in various update sources or to determine the update history of particular street centerlines in TIGER.

The panel believes that the Census Bureau will be best served by an incremental development approach in redesigning the MAF/TIGER database—that is, that modernization should not be attempted on the entire database structure at once but rather divided into smaller, achievable subtasks. Each subtask would then be carried out—and rigorously tested—in turn. A major goal in approaching the work in this manner is to have available at all times a database structure for MAF and TIGER that is operable and capable of achieving all of its census missions. That system will be a hybrid, gradually evolving into the completed new design as work increments are completed. A hybrid that is continually operable and capable is preferable to the development and implementation en masse of a completely new database structure, which could easily be jeopardized and rendered a complete loss by changes in budget or resources.

As we noted in Section 6–B.3, we are encouraged by the designation of a system architect for the MAF/TIGER database redesign, and strongly urge that this person's work be done in conjunction with a system architect for the decennial census as a whole. That said, we are concerned about aspects of the relationship between the Census Bureau and its principal Objective Two contractor, since our understanding is that a contractor employee will serve in the system architect role. In panel members' experience with projects of this type, it is unwise to assign that much design, management, and decision authority in the hands of a contractor: a system architect should have a deep understanding of the existing and proposed systems, their history, and their interconnections and relations to other parts of the enterprise and, typically, contractors do not have that depth of institutional knowledge. We believe that the Census Bureau should have a Bureau employee paired with contractor personnel for every key task and skill, especially for senior management and decisionmaking roles. Bureau staff should certainly learn from contractors, but they should not be dependent on them in the long term for skills or knowledge. Key decisions should be made by Census Bureau staff.

We note that making the transition from the existing TIGER database structure to Oracle-based systems will necessarily mean a switch to object-oriented programming, design, and testing. In our experience with computer science projects, the first-time adoption of object-oriented programming approaches is a particularly tricky one, fraught with unanticipated difficulties and surprises. Consistent with recommendations we make in the next section regarding software engineering approaches, we strongly encourage the Bureau to develop a small review team of experienced computer scientists and software developers to monitor and facilitate the Bureau's move into this new development paradigm.

Our final comment on the MAF/TIGER database redesign is that we believe it could present a unique opportunity to build ties to the software development community. Development of the original TIGER in the mid-1980s helped spawn the geographic information systems industry. We believe that the new database structure—and the attendant rewriting of the support software currently used to update TIGER, create maps, and match addresses to geographic coordinates—gives the Census Bureau the chance, at little cost, to again influence the industry if it pursues the redesign with a measure of openness. By publishing the technical description and offering public access to the code (but not, obviously, the complete and Title 13-protected MAF and TIGER data) of its support software, developers would have the chance to scrutinize, modify, mimic, and improve the software. Those developer contributions could, in turn, be adopted or rejected by the Bureau as it pursues its own development, but at least partners in the broader community will have had the opportunity to participate and contribute.

> ***Recommendation 6.3:*** **As part of the MAF/TIGER redesign, the Census Bureau should consider ways to make its application code for mapping, geocoding, digital exchange, map editing, and other functions openly available in order to facilitate continued ties to and improvement in geographic information systems software applications and to tap the feedback of the broader computer science/software development community.**

6–D CHALLENGES IN TRANSITION FROM LOGICAL TO PHYSICAL INFRASTRUCTURE

A business process or logical architecture model will define the activities and the informational interfaces and dependencies required to carry out the 2010 census. Between now and the dress rehearsal in 2008 (with an opportunity to do related testing in 2006), an integrated information system—a physical technical infrastructure—must be put into place to support those activities and satisfy their informational requirements. In preparation for the refinement of the 2010 logical architecture and the transition to a physical infrastructure, we offer some further comments based on past experience with reconfiguring information systems. We raise these points—some of them cautionary in nature—not to deter the Census Bureau from proceeding with architecture modeling efforts but to emphasize the difficulty and importance of the task.

6–D.1 Potential Pitfall: Locking in Physical Infrastructure Too Early

A major danger in making the transition from retooled logical infrastructure to completed physical infrastructure is a "rush to judgment"—a decision to finalize physical structures too early. Moore's Law—the adage that computing power tends to double roughly every 18 months—is well known; the rate of change in the computer technology world is indeed astounding. Thus, in settling on the purchase of a particular computer or software package, the Census Bureau runs the same risk faced by millions of personal computer buyers in the past several years: namely, nearly instant obsolescence, as the capabilities of the chosen product are bested shortly thereafter by the next generation of product.

As discussed further in Section 5–A, the selection of PCDs is a particular area where the Census Bureau should remain cognizant of the dangers of deciding on physical form too early. At present, small-scale tests of basic skills are being conducted—navigation using a map displayed on a palm-sized screen, administration of a computerized questionnaire on a small computing device, and so forth. It is important that the Census Bureau continue to conduct prototype testing of this nature in order to get some sense of current capabilities. However, it is likely to be a mistake to draw final conclusions on qualities like desired PCD weight, size, and memory capacity based on early test results. PCDs are relatively simple computing devices with reliable storage and test input facilities. Additional features that may be desired include: a color display with good resolution, a GPS latitude-longitude acquisition device, electronic communication facilities such as a landline modem, and perhaps encryption and decryption capabilities. However, the most important product of early PCD testing is not so much a checklist of desired features but a clearly articulated plan of the workflows and information flows that must be satisfied by PCDs, as they fit into the broader technical infrastructure of the census.

6–D.2 Enterprise Architecture as Learning Tool and Guide to Organizational Change

The end goal of business process or logical architecture reengineering is the production of a smoothly functioning finished physical architecture—an amalgam of people, software, computer systems, and telecommunications systems. Given this purpose, it is perhaps too easy to cast the effort as purely technical and technological, but this would be a highly inaccurate impression. We strongly encourage the Census Bureau to take full advantage of the exercise of architecture reengineering by viewing the effort not merely as the means to reengineer the Bureau's computer systems but also as a key information tool to reengineer its own organization and operations.

As indicated above, IDEF0 logical architecture models emphasize function and process independent of labor and departmental boundaries within an organization. Large organizations that develop rigid internal divisions over time can benefit from—and find refreshing—the exercise of stepping back and specifying basic flows of information, without the need to consider which division performs a given function or to which directorate it may report. For the Census Bureau, this logical architecture modeling represents a "new, and very different, perspective on decennial census operations," one "based on logical groupings of functions [and highlighting] the commonality across similar processes that were developed independently for different operations" (Centech Group, Inc., 2002a:vii). Accordingly, this new approach represents a potential step away from the "compartmentalized thinking" the panel warned against in its letter report (National Research Council, 2001c).

By these comments, we do not suggest the need for wholesale change in the way the Census Bureau is currently structured. What we do suggest is that the Bureau could benefit greatly from the development of a task-based project management approach. The analysis of information flows in architecture models may suggest logical clusterings of activities—or redundancy in activities—and provide clues for how parts of the Bureau may best be mobilized to carry out each task.

6–D.3 Changing Architecture and Methods Simultaneously

Reengineering the Census Bureau's information systems is a very large and complex project in its own right. However, it is made vastly more difficult because the Bureau will be reengineering a very large and complex integrated system *at the same time* as it attempts to make substantial changes in the tools and methods it plans to use—for instance, the migration of the MAF/TIGER system to a commercial off-the-shelf database system, the development (in the ACS) of a complete data system parallel to the census, and the implementation of new response modes. The added difficulty involved in developing new methods simultaneously with new architecture argues ever more strongly for a strong, coordinated system architect for the census, as synchronized efforts will be key to successful implementation.

6–D.4 Improving Software Engineering and Development

The Census Bureau has indicated that, as it pursues the TIGER database modernization, it has also taken on the goal of improving the Bureau's Capability Maturity Model (CMM) score, a measure of an organization's maturity in software engineering (Franz, 2002). This is certainly a worthwhile goal, but one that we caution should not be approached casually. In isolation, taken as a single goal, experience suggests that organizations take approximately 2–3 years to move up one CMM level. The fact that the Census Bureau is simultaneously undertaking broader systems engineering and major technology projects in the TIGER redesign and PCD implementation may further extend the time needed to increase the score or to complete the systems projects under development.

As with our other cautionary notes in this chapter, we raise the difficulty of the task not to discourage the Census Bureau from taking action but rather to state that it is more complicated and time-consuming than may be expected. Allowing one of these paths—improving software engineering capability or designing system architecture—to proceed in isolation from the other could be a critical and costly error if time and resources elapse without both contributing jointly to census objectives.

> ***Recommendation 6.4:*** **The Census Bureau should generally improve its software engineering processes and should pursue its goal of raising its Capability Maturity Model score in software development. In particular, the Bureau should focus on available tools and techniques in rigorously developing and**

tracking software requirements and specifications. In beginning the task of improving its software practices, though, the Census Bureau must recognize that the effort is a difficult one, requiring high-level commitment in the same manner as architecture reengineering.

On a related note, and consistent with the Bureau's broader efforts to improve software engineering practices, we urge the Census Bureau to assess its standards and planning assumptions related to hardware and software experience. It is well known in the software development community that it is vastly more expensive to detect bugs and operational errors when hardware and software have been fielded than to catch those bugs during prerelease testing; this lesson has also been learned by other survey organizations as they have moved increasingly into computer-assisted interviewing methods (National Research Council, 2003b). For the census—a technologically intensive survey of grand scale with a strict timeline—catching software errors early is particularly important to smooth operations.

Recommendation 6.5: **The Census Bureau should evaluate and improve its protocols for hardware and software testing, drawing on expertise from the computer science and software development communities. Rigorous hardware and software testing should be factored into census operational schedules, in addition to the field testing performed in the 2006 proof-of-concept test, the 2008 dress rehearsal, or such other formal census tests as may arise.**

CHAPTER 7

Coverage Measurement

IT IS ALMOST CERTAIN that the resident population of the United States on April 1, 2000, was not exactly equal to 281,421,906, even though that is the total reported by the 2000 census. No decennial census has ever attained a perfect, complete count of the population; the results of a census represent the best effort to count every resident once and only once, but some people are inevitably missed in the count and others are counted multiple times. The possibility of undercount in the census has been a longstanding concern, particularly since the level of undercount has been estimated to vary differentially by racial and ethnic groups in recent censuses. In the 2000 census, follow-up research eventually concluded that the 2000 census may have experienced a net *overcount,* the first such occurrence in census history. Given the inherent complexity of the decennial census task, it is crucial that the census include programs that permit examination of the accuracy and completeness of the count; development of such a coverage measurement plan remains a major challenge in planning the 2010 census.

In this chapter, we outline our suggestions for the shape of a coverage program in 2010 relative to that used in 2000 (Section 7–A). We then comment on demographic analysis, an alternative coverage measurement methodology that provided a very useful point of comparison in the 2000 census coverage measurement program (7–B). Finally, we discuss the use of administrative records, the focus of a major experiment in 2000 (7–C).

7–A THE SHAPE OF COVERAGE MEASUREMENT IN 2010

The quality of census coverage and the possibility of statistically adjusting census totals to reflect coverage gaps developed into the defining issues of the 1990 and, especially, the 2000 censuses. That some people are missed in the census count while others may be multiply counted is virtually inevitable and has never been in dispute, even since the earliest censuses. However, the intensity of the political debate over census coverage, over the differential nature of census undercount by race and other demographic groups, and over the reliability and validity of statistical adjustment grew enormously in the past two censuses, to the point that the 2000 census was conducted under an unprecedented level of oversight and suspicion.

The results of 2000 coverage evaluation efforts have not settled the ongoing debate over census adjustment. In the 2000 census cycle, the Census Bureau faced three separate points at which a decision on statistical adjustment had to be rendered: March 2001 for redistricting purposes, October 2001 for federal fund allocation and other purposes, and March 2003 for use as the base for postcensal population estimates. In all three instances, the Bureau opted against adjustment as results of the Accuracy and Coverage Evaluation (ACE) Program—the follow-up survey used to assess census coverage and derive adjustment factors—showed unanticipated results. In March 2001, concern over the discrepancy between ACE-adjusted census counts and the alternative population count derived through demographic analysis was sufficiently large to deter adjustment; ACE research through October 2001 resolved some conceptual issues and led to a significantly lower estimate of national net undercount in the census, but still left too many unanswered questions for the Bureau to recommend adjustment. By March 2003, Bureau reexamination of the ACE (ACE Revision II, in their terminology) suggested a national net *overcount* of population, the first such finding in census history (although different racial and demographic groups still experienced significant net undercount at the national level).

As Singh and Bell (2003) noted at the panel's September 2003 meeting, the Census Bureau's plans for coverage measurement in 2010 will be driven by the following general goals: (1) to produce measures of the components of coverage error, (2) to produce these measures for demographic groups, geographic areas, and key census operations, and (3) to provide measures of net coverage error. However, to our knowledge, the Bureau has not yet made concrete plans for testing improved coverage measurement procedures.

Our approach in this report is primarily pragmatic. We believe it is vitally important that the 2010 census include mechanisms that permit in-depth evaluation of how well the census performs in enumerating the population, both as a whole and differentially for population subgroups. Most simply, a program is necessary for the measurement of coverage although it need not be a census "coverage measurement" program as that term has come to be known. However the 2010 coverage measurement program is structured, it is essential that it be addressed early, that it be the subject of research and evaluation throughout the years leading up to 2010, and that it be included in the 2006 proof-of-concept test and the 2008 census dress rehearsal.

It is not necessary that the coverage measurement program for the 2010 census follow the exact same structure and script as the 2000 Accuracy and Coverage Evaluation Program. Indeed, in light of the analysis of the Panel to Review the 2000 Census (National Research Council, 2004), repetition of the 2000 ACE in 2010 without substantial improvement would be detrimental; it would likewise be harmful if the 2000 methodology had to be used, as is, as a fall-back position absent research and resolution of a plan in the years preceding 2010.

To the extent that coverage measurement in 2010 makes use of a postenumeration survey (PES) combined with dual-systems estimation (DSE)—the primary approach used in the past two censuses—we have made several suggestions in this report that could improve the methodology. These include:

- further research on matching census records, searching the nation and matching by name and date of birth, as part of the unduplication effort (Section 5–E);

- inclusion of the group quarters population in the postenumeration survey (and, more generally, reconciliation of group quarters enumeration with housing unit enumeration) (Section 5–B.2); and
- better definition of census residence rules and better communication of the same to respondents through redesigned questionnaires and CAPI techniques (Section 5–B.3).

In addition, the panel hopes that an improved MAF/TIGER will contribute to a reduction in geocoding errors, which proved to be a point of concern in the 2000 ACE. But, beyond those steps, we do not believe it appropriate to delve into the mechanics of the PES-DSE combination in this report nor to offer specific recommendations on how it should or should not be implemented in 2010, given that our active discussion with the Bureau on those possibilities began very late in the panel's term.

In the balance of this chapter, we discuss demographic analysis in Section 7–B. Used as a coverage measurement tool since the 1950 census, demographic analysis remains an important approach that could benefit from some strategies for improvement. Finally, while research on administrative records remains, to a great extent, a topic for experimentation rather than implementation in 2010, such research could feed into coverage evaluation efforts (7–C).

7–B ENHANCING DEMOGRAPHIC ANALYSIS FOR 2010

The success of demographic analysis as a tool for census coverage evaluation depends on access to accurate and highly reliable information. It has been generally assumed that the data used for demographic analysis were of sufficient quality to support highly accurate intercensal annual estimates of the size of the United States population, and that these estimates could be accumulated over the decade to obtain a figure against which the next census enumeration could be benchmarked. Demographic analysis has also been an important tool for assessing the black-white differential undercount.

Demographic analysis requires highly accurate data on three components of population change: fertility, mortality, and net migration. In the case of fertility data, the surveys done in the 1960s to determine the completeness of the data since 1930 have generally confirmed the accuracy of these data, both in terms of the numbers of births reported and the characteristics of mothers and offspring. However, alternative assumptions of the completeness of the data were utilized in the application of demographic analysis to the coverage of the 2000 census (Robinson, 2001).

For mortality data, while estimates of the numbers of deaths reported can be assumed to be relatively accurate, studies of infant mortality (Hahn, 1992; Hahn et al., 1992) have raised serious questions about the accuracy of reports about the racial background of the deceased. Coroners and medical examiners, for example, are not always able to reliably determine the race of the deceased in the absence of information from the decedent's family.

Immigration and emigration data are undoubtedly the most problematic component for demographic analysis. Immigration estimates consist of two subcomponents: documented and undocumented immigrants. For demographic analysis, the Immigration and Naturalization Service (INS)[1] has been an important source of information about United States immigration for documented and undocumented arrivals. While estimates of immigrants are relatively complete when people arrive with the proper documentation, estimates of undocumented immigration are no more than educated guesses based on INS arrests and deportations and on imaginative use of census and household survey data. For the 2002 intercensal estimates, the Census Bureau was able, for the first time, to incorporate data from the 2000 and 2001 Census Supplementary Surveys. These data were corroborated with data from the INS and were found to be reasonably consistent for documented immigrants. However, as in the past, undocumented immigration continues to be treated as a residual category resting on the unsubstantiated assumption that the cov-

[1] In March 2003, authority that had been vested in the INS was divided among three bureaus in the newly-formed Department of Homeland Security—the U.S. Citizenship and Immigration Services, the Bureau of Immigration and Customs Enforcement, and the Bureau of Customs and Border Protection.

erage of undocumented immigrants is not significantly different from that of immigrants who arrive with proper documentation.

In the past, a great deal of research by federal agencies and others has supported the view that most undocumented entries arrive from Mexico and Central and South America. While the majority of undocumented entries may indeed arrive from these regions, an ever larger number may be arriving from other troubled parts of the world such as Asia and Africa. Because of the relatively porous border with Canada, undocumented immigrants from Asia, Africa, and even Europe are almost certain to prefer entering the United States from the north. Another option is to enter the country with a tourist visa and then simply remain beyond the visa's expiration. Given the United States' increasingly global ties and the comparatively liberal rules for entry into Canada, it may no longer be reasonable to assume that undocumented immigration from the north is negligible. Research about the extent of undocumented immigration from Canada is very limited, and additional research could be warranted. Undoubtedly, the transfer of portions of INS to the new Department of Homeland Security is likely to result in these issues being given a higher priority than in the past. However, it is not clear how well the size of this population can be estimated even with maximal resources.

Demographic analysis is also important because it provides reasonably unbiased national estimates of the number of native-born black Americans and because it supports the production of intercensal estimates for subnational areas, though estimates of interstate migration are also needed. ACS data might support substantial enhancement of the current approach to estimating internal migration, which uses tax return data. However, as discussed in Chapter 4, much research is needed to determine how to exploit the ACS data for this purpose.

Like National Research Council (2004), we believe that demographic analysis has proved to be a useful independent benchmark—but is not in itself a gold standard—for assessment of census coverage. Particularly if estimates of immigration and emigration can be improved, we believe that it should continue to play a valuable role in coverage measurement in 2010. In addition, we noted earlier that the fundamental research on the completeness of birth registration needs to be updated and that questions have been raised regarding the racial reporting in death data. Accordingly, the demographic analysis program for 2010 would benefit strongly from renewal and revisiting of research on the basic assumptions of the methodology. Accordingly, we endorse a recommendation from that report, with modification (National Research Council, 2004:Rec. 6.2):

> ***Recommendation 7.1:*** **The Census Bureau should continue to pursue methods of improving demographic analysis estimates, working in concert with other statistical agencies that use and provide data inputs to the postcensal population estimates. Work should focus especially on improving estimates of net immigration. Attention should also be paid to quantifying and reporting uncertainty in demographic estimates. Updated assessments of the assumptions underlying demographic analysis estimates, including the completeness of birth registration, should also be considered.**

7–C ENHANCING ADMINISTRATIVE RECORDS ANALYSIS FOR 2010

For several years, the possibility of a census conducted in part (or even in whole) by use of administrative records—the person-level data maintained by a host of federal government programs—has been the focus of recurring discussions. The potential applicability of administrative records in the census has increased, both as the administrative records databases maintained by the government have become more complete and as the computing capacity to merge and manage multiple lists has become more powerful and more sophisticated.

7–C.1 Administrative Records Experiment in the 2000 Census

As part of the program of experiments planned to accompany the 2000 census, the Census Bureau initiated an administrative records experiment, which came to be known as AREX 2000.

The experiment examined the possibility of using administrative records for several purposes, including the derivation of population estimates (that is, an administrative records census). Other uses that AREX 2000 was intended to investigate include the use of information for improvement of the Master Address File, help with census unduplication, and refinement of intercensal or postcensal population estimates. The experiment also considered the possible use of administrative records as a resource for imputation, either for households with no report at all or for those with missing data items. Results of the experiment are reported in Bauder and Judson (2003); Berning (2003); Berning and Cook (2003); Heimovitz (2003); and Judson and Bye (2003).

AREX 2000 was limited in scope to five county-level sites: the city of Baltimore, Maryland, the surrounding Baltimore County, and Douglas, El Paso, and Jefferson Counties, Colorado. The 2000 census population of the test sites is 2.6 million, in 1.2 million households. AREX 2000 was charged with using administrative records to provide population counts and demographic characteristics for census tracts and blocks in the sites.

Though analysis was limited to the test sites, AREX 2000 used as its base a national-level resource: the Census Bureau-compiled Statistical Administrative Records System (StARS). StARS merged and unduplicated records from six major administrative records databases:

- Internal Revenue Service (IRS) Individual Master File (IMF 1040),
- IRS Information Returns Master File (IRMF W-2/1099),
- Department of Housing and Urban Development (HUD) Tenant Rental Assistance Certification System (TRACS) File,
- Center for Medicare and Medicaid Services (CMS) Medicare Enrollment Database (MEDB) File,
- Indian Health Services (IHS) Patient Registration System File, and
- Selective Service System (SSS) Registration File.

Demographic data to impute missing values remaining from these sources were derived from the Census Numident file, an edited version of the Social Security Administration's master file of assigned Social Security numbers (also known as the Numerical Identification, or Numident, file).

The specific data assembled to form StARS were from 1999 (the IRS data sources were for tax year 1998). These resources were used as both records of individuals and listings of addresses. The information culled from the files was selected based on its currency and perceived quality; records were also assessed by whether they could be geocoded (that is, whether the addresses could be matched to the TIGER geographic database).

With respect to data on individuals, 875 million records were initially available after merging. After unduplication and removal of known deceased individuals and persons residing outside the United States, a file of 257 million individuals was produced (Judson and Bye, 2003:11). With respect to addresses, almost 800 million were available at the start, and after unduplication and removal of business addresses (and other operations), approximately 147 million addresses were produced, of which 73 percent were able to be geocoded (Judson and Bye, 2003:15).

Two methods were examined for taking an administrative records census, referred to as "top-down" and "bottom-up." The top-down approach was a raw administrative-records-only census: tallying the number of people on the StARS (merged and unduplicated file) with addresses geocoded to the test site locations. The bottom-up approach matched the StARS address list with the Decennial Master Address File (DMAF) in order to simulate the mailout/mailback and nonresponse follow-up phases of the census.[2] Data from StARS records with addresses matching to the DMAF are thought of as the mailout/mailback piece; to simulate nonresponse follow-up, 2000 census counts for DMAF addresses not found in StARS were added to the "mailout/mailback" administrative records count.[3]

---

[2]The DMAF is the version of the MAF that is extracted prior to the census and used to print mailing address labels and monitor mail response.

[3]An administrative records census with a bottom-up design would typically have field follow-up for DMAF addresses that are not found in the administrative records database. In lieu of actually doing field follow-up as part of AREX 2000 (and incurring substantial costs as a result), 2000 census counts were used for the addresses that would have been designated for field follow-up.

AREX 2000 had important limitations, which are acknowledged by the Bureau in the reports of the experiment. First, AREX 2000 used a version of StARS created using 1998 and 1999 data, creating a time gap relative to the target census reference data of April 1, 2000. Second, additional structure in the experimental design of AREX 2000 could have provided more information concerning the value of various components of the AREX 2000 operation. Specifically, evaluation of the choice of "best" address (as well as other field and clerical operations) could have been carried out using a more elaborate design. The panel's first interim report made suggestions that were not included in the AREX 2000 design, including (1) integration of AREX 2000 with the Accuracy and Coverage Evaluation so that it could be determined which households and individuals ACE tended to miss, and (2) field follow-up to help evaluate the quality of the merged administrative records list. Another limitation is that, for the purposes of the experiment, the Bureau elected not to consider administrative records from commercial sources or to try to draw from state and local government records.

Nevertheless, even with the various limitations, AREX 2000 was a valuable experiment. It demonstrated the feasibility of merge and unduplication operations that had not been evaluated previously. AREX 2000 also provided extremely useful information on the value of administrative records for use in assisting nonresponse follow-up. In evaluating the experiment, the Census Bureau concluded that the top-down approach to an administrative records census experienced an 8 percent undercount across the test sites, a substantial figure. However, this undercount was cut to 1 percent by the bottom-up procedure (Judson and Bye, 2003). These undercounts were carefully examined by various demographic characteristics and at different levels of geographic aggregation. In addition, logistic regression models were used to help predict for which types of households administrative records data might be useful for providing whole-household imputations.

7–C.2 Administrative Records for 2010

AREX 2000 demonstrated the potential for the use of administrative records in the census process. Exactly how great that potential is—or, put another way, how close these methods are to actual implementation in the census—is less clear. At the very least, administrative records research should be pursued for further and fuller experimentation as part of the 2010 census. That said, the possibility of a substantial role—for instance, use of administrative records for help in nonresponse follow-up, imputation, or targeting MAF improvements—ought not be rejected out of hand. Much work would be needed to develop and implement any of these ideas in 2010; should the decision be made to try to use administrative records in the census, it would be important to focus on one or two applications at most and to include an evaluation of those applications in the 2006 census test.

7–C.3 Other Possibilities: Megalist and Reverse Record Check

From the conceptual standpoint, at least two other possibilities could be posited for more extensive use of administrative records in the census context. These are sufficiently promising as to warrant additional research and experimentation, though they are admittedly more likely to be part of a discussion of methodology in 2020 rather than 2010.

The first possibility is *megalist*, which is the concept of continuing to develop merged lists of administrative records as an independent listing of the population (National Research Council, 1985; Ericksen and Kadane, 1983). Thus, the administrative list could be used as the second component of dual-systems estimation (filling the role of the postenumeration survey in the 2000 Accuracy and Coverage Evaluation, for instance) or as a third component in triple-systems estimation. The advantages of a megalist over a postenumeration survey are the cost savings on field data collection and the possibility of improved representation of hard-to-enumerate populations (since relevant administrative lists could be used, e.g., from welfare program participants or the Indian Health Service records used in the current StARS). The primary advantage of triple-systems estimation is a reduced reliance on the independence assumption used in dual-systems estimation. The disadvantages of the megalist approach include questions of the representativeness of the merged lists, the quality of residence location information, and the availability of reliable information for matching the separate lists (e.g., date of birth). This latter concern is especially important for triple-systems estimation, where the amount of matching is tripled in comparison to dual-systems estimation.

The second possibility is *reverse record check*, the primary method used by Statistics Canada in evaluating the coverage of the Canadian census (Gosselin, 1980). In this technique, samples of births, immigrants, those counted in the most recent census, and those missed in the most recent census (which is roughly provided by the previous implementation of this program) are each traced to their current address to arrive at a target count for each area to compare against the census counts. Since the Canadian census is taken every 5 years, tracing addresses forward in time requires finding people after a lag of only 5 years. On the other hand, for the United States census, tracing would have to extend over a lag of 10 years. This crucial difference in the application of this technique to the United States census was tested by the Census Bureau in 1984 in the Forward Trace Study (Hogan, 1983), in which it was discovered that tracing over 10 years was not feasible. However, administrative lists are of higher quality than in 1984, and it may be that reverse record check should be reevaluated as a possibility in the United States census.

Part III

The Census Bureau's Research and Research Design

CHAPTER 8

Evaluations

RECENT DECENNIAL CENSUSES HAVE INCLUDED a program of planned evaluation studies that focus on the quality of census operations and the data that result from them. In some respects, the 2000 census could be interpreted as having two such programs. One was the series of detailed reports from the Accuracy and Coverage Evaluation (ACE) program. Numerous supporting reports accompanied each of the decision documents on census adjustment issued by the Bureau's Executive Steering Committee for ACE Policy (ESCAP) in March and October 2001, and another series of supporting reports accompanied the final estimates from the 2000 coverage evaluation program—called ACE Revision II—when those estimates were released at a joint meeting of our panel and the Panel to Review the 2000 Census in March 2003.

The second evaluation program of the 2000 census was the planned slate of evaluations of census operations and of the quality of the census content, which was administered by the Census Bureau's Planning, Research, and Evaluation Division (PRED). The original plan of PRED evaluations for 2000 was very ambitious, including 149 separate studies; 18 of these were "cancelled" only in the sense that they were instead expedited and completed as part of the ESCAP evaluation series. But subsequently, in at least two major waves (in early and late 2002), the evaluation program "was refined and priorities reassessed due to resource constraints at the Census Bureau" (U.S. Census Bureau, 2003d). In an "[attempt] to obtain the best balance of resources" between "completing and releasing Census 2000 data products" and "conducting key Census 2000 evaluations" (U.S. Census Bureau, 2003d), the Census Bureau ultimately reduced the list of studies from 149 to a still-formidable 91. In addition, a series of 15 topic reports was developed based on groupings of evaluation reports; the Bureau released the individual evaluation reports only after the relevant topic report was publicly released.

In this chapter, we discuss suggestions for developing the evaluation program for the 2010 census. In Section 8–A, we outline major challenges that we perceive in defining evaluation studies for the 2010 census and, more broadly, redefining the research and evaluation program of the Census Bureau. In Section 8–B, we describe the Master Trace Sample, an evaluation tool that we—like other National Research Council panels—believe may be particularly critical to learning about census operations and guiding future practice.

8–A STRENGTHENING THE EVALUATION PROGRAM OF THE 2010 CENSUS

The staff of both our panel and the Panel to Review the 2000 Census received access to the Census Bureau's PRED-series evaluations and topic reports on an advance release basis, for which we thank the Census Bureau (the General Accounting Office, Department of Commerce Inspector General, and Congressional staff received the evaluation reports on the same basis). While our arguments in this report reflect observations from the evaluation reports, we do not here attempt a comprehensive review of the entire slate of Census Bureau 2000 census evaluations. This decision reflects a change in the panel's charge described in Chapter 1 to a more forward-looking study of the developing plans for the 2010 census.

Of the Census Bureau's 2000 census PRED-series evaluations, the Panel to Review the 2000 Census commented (National Research Council, 2004:Sec. 9–B):

> We applaud the effort the Census Bureau devoted to evaluation of the 2000 census. Yet we must note the serious deficiencies of many (but by no means all) of the evaluation studies released to date. Too often, the evaluations do not clearly answer the needs of relevant audiences, which include 2000 census data users who are interested in data quality issues that bear on their analyses and Census Bureau staff and others who are concerned with the lessons from the 2000 experience that can inform 2010 planning. No single evaluation will necessarily speak to both audiences, but every evaluation should clearly speak to at least one of them.
>
> Yet many of the completed evaluations are accounting-type documents rather than full-fledged evaluations. They provide authoritative information on such aspects as number of mail returns by day, complete-count item nonresponse and imputation rates by type of form and data collection mode, and enumerations completed in various types of special operations (e.g., urban update/leave, list/enumerate). This information is valuable but limited. Many reports have no analysis as such, other than simple one-way and two-way tabulations. Reports sometimes use different definitions of variables, such as type of form (self or enumerator), and obtain data from files at different stages of processing, with little attempt to reconcile such differences. Almost no reports provide tables or other analyses that look at operations and data quality for geographic areas.

Based on our reading of many of the evaluation reports, we concur. Like the Panel to Review the 2000 Census, we believe that there is merit in the completed evaluation studies of the 2000 census as rough documentation of operational processes. We also appreciate the difficulty faced by evaluators in marshaling data resources that could be daunting in scale or, worse, simply not amenable to necessary evaluation work (the dominant example being the MAF Extract described in Chapter 3, in which the logical flags associated with different address updating operations were not conducive to tracking the complete history or original source of an address).

That said, we believe that the 2000 census evaluation work is suggestive of two broader problems. We hope that by addressing these problems the Bureau is able not only to strengthen the evaluation program for 2010 but also to redefine the way it thinks about research.

8–A.1 Correcting the Disconnect Between Research and Operations

The first broad problem is, as we perceive it, a real disconnect between the Bureau's research objectives and its operational and planning groups. In our first interim report (National Research Council, 2000a:28–29), we recommended that the Census Bureau "develop a detailed plan for each evaluation study on how to analyze the data collected and how to use the results in decision making." Specifically, we suggested that these plans "include detailed information on how the data will be analyzed, how the results obtained will inform decisions about the 2010 census design, and what resources, in terms of data collection costs and staff expertise, are required." We continue to believe that a strong connection between a research base and operational decisions is vital. However, we see signs in many aspects of the emerging 2010 census plan that research is not strongly connected to planning needs and goals.

The clearest such sign is the articulation of the basic "three-legged stool" approach to the 2010 census itself. As noted in Section 2–C.2, the key initiatives of the 2010 census plans were developed as the 2000 census was being collected and tabulated. As such, they were developed without the benefit of completed evaluation studies and research. Thus, the MAF/TIGER Enhancements Program was pursued as an objective without reference to either the unique contribution of address update sources to the Master Address File or detailed assessment of geocoding and enumerator difficulties associated with TIGER database inaccuracy. Likewise, the American Community Survey was proposed before all information was known about nonresponse—in full and item by item—to the 2000 census long form. As elsewhere in this report, we emphasize our agreement with the basic goals of the 2010 census plan and, as we said in Section 2–C.2, it was appropriate that major initiatives for the 2010 census were proposed and initiated as early as they were. Still, it is significant that these initiatives did not derive naturally from the results of

2000 census evaluations or the problems detected and measured by those evaluations.

In other areas, the Census Bureau's planning and research entities operate at either a high level of focus (for decision making) or at a microlevel that tends toward detailed accounting of individual census processes (as in the series of 2000 census evaluations), with little to bridge the gap. Examples of decisions not well informed by research include:

- the decision to implement a complete block canvass for 2010 address list updating, without full research into alternative methods or definitive knowledge of the unique contributions of address update operations to the 2000 MAF;
- the decision to favor imputation methods over extended nonresponse follow-up operations in the 2000 census, without complete research into the effect of imputation on resulting data series or development of modern imputation tools; and
- the decision to limit the number of persons who can respond on the basic census return form from 7 to 6 in 2000, which served the need to shorten the form but may have had unintended consequences for the reporting of large households.

As it designs its research and evaluation program for 2010, the Census Bureau should work to bridge the gap between research and operations in the census process; evaluations should be forward-looking and designed to inform and satisfy specific planning objectives. The goal should be research studies that produce real data that lead to actionable results.

Accordingly, we offer a general recommendation for the development of a research and evaluation plan for the 2010 census. This recommendation represents an endorsement and restatement of a similar recommendation from the Panel to Review the 2000 Census (National Research Council, 2004:Rec. 9.2).

> ***Recommendation 8.1:*** **The Census Bureau should materially strengthen the evaluation component of the 2010 census, including the ongoing testing program for 2010. Plans for census evaluation studies**

should include clear articulation of each study's relevance to overall census goals and objectives; connections between research findings and operational decisions should be made clear. The evaluation studies must be less focused on documentation and accounting of processes and more on exploratory and confirmatory research while still clearly documenting data quality. To this end, the 2010 census evaluation program should:

1. **identify important areas for evaluations (in terms of both 2010 census operations and 2020 census planning) to meet the needs of users and census planners and set evaluation priorities accordingly;**
2. **design and document data collection and processing systems so that information can be readily extracted to support timely, useful evaluation studies;**
3. **focus on analysis, including use of graphical and other exploratory data analysis tools to identify patterns (e.g., mail return rates, imputation rates) for geographic areas and population groups that may suggest reasons for variations in data quality and ways to improve quality (such tools could also be useful in managing census operations);**
4. **consider ways to incorporate real-time evaluation during the conduct of the census;**
5. **give priority to development of technical staff resources for research, testing, and evaluation; and**
6. **share preliminary analyses with outside researchers for critical assessment and feedback.**

8–A.2 Pursuing New Research Directions

The second broad problem that we believe is suggested by the evaluations of the 2000 census concerns the relative lack of attention to some types of research, key among them the investigation of opportunities for targeting and efficiency. The Census Bureau's 2000 census evaluations are, apparently by design, national in scope and focus; there is very little disaggregation by geography or (in some cases) demographic group that might shed light on important local variation in census operations. Even in those cases where disaggregation is attempted, the results can be confusing and unhelpful for planning. Perhaps most notably, evaluation reports on the Local Update of Census Addresses (LUCA) program offer tabulations of additions and deletions broken down by state, even though it was smaller-area governments (counties, minor civil divisions, places, and reservations) that participated in LUCA; the breakdown by state provides some insight but, generally, misses the important story of participation and nonparticipation in the LUCA process (Owens, 2003, 2002).

Block canvassing, group quarters enumeration, small multiunit structures, and rural areas—as well as other topics raised throughout this report—are cases in which 2010 census planning would benefit by departing from the "one size fits all" approach that often characterizes census operations. Just as it is likely that canvassing for addresses in selected areas may be effective relative to a blanket block canvass, so too is it likely that the accuracy of the count of special hard-to-count populations may be improved by tailoring questionnaires and enumeration methodologies to reach them. Accordingly, we recommend:

> ***Recommendation 8.2:* A major focus of the Census Bureau's ongoing research and evaluation program must be opportunities for targeting and efficiency—tailoring approaches to key population groups and areas rather than pursuing a "one-size-fits-all" approach.**

Our discussion in this report also suggests areas where increased focus on cognitive testing and questionnaire design would be beneficial. Better articulation and presentation of the census residence rules could help identify or deter person duplication in the census. Moreover, the establishment of a parallel data system in the ACS highlights the importance of maintaining appropriate consistency in questionnaire content and design; the divergent residence rules for the ACS and the census stand as an open question that should be resolved, and wording and structuring of race and Hispanic origin questions should be consistent between the two questionnaires. On a related matter, human factors and usability testing should become increasingly important in the Bureau's research and evaluation programs, due to the plans to deploy portable computing devices among the large corps of temporary census enumerators and the wider availability of the self-response Internet option.

8–A.3 Exploiting Existing Data Resources

While it may be tempting to look at the completed 2000 census evaluation reports and topic reports and conclude that evaluation of the 2000 census is complete, the panel argues that much remains to be learned from the extant operational data from the 2000 census. Further disaggregation and mining of these data should be an informative and relatively inexpensive way to formulate a stronger research base for the 2010 census and its constituent programs. We recommend the following (see also National Research Council, 2004:Rec. 9.1):

> ***Recommendation 8.3:*** **The Census Bureau must mine and fully exploit data resources currently available in order to build a research base for the 2010 census and to further evaluate the 2000 census. These resources include:**
>
> - **microdata from the 2000 Accuracy and Coverage Evaluation and its related Person Duplication Studies;**
> - **extracts from the Master Address File;**
> - **the Local Census Office Profile dataset;[1]**

[1] Created as part of the 2000 census evaluations program, the Local Census Office Profile (Imel, 2003) is an extract from a merged data set drawing from, among other sources, the Decennial Master Address File and the Hundred Percent Unedited and Edited Files (the person-level files of census reports that are tabulated to produce census results).

- **a match of census records and the March 2000 Current Population Survey; and**
- **the Master Trace Sample.**

We address one of these as-yet-untapped data resources—the Master Trace Sample—in detail in the following section.

8–B MASTER TRACE SAMPLE

The idea for what has come to be known as the Master Trace Sample (MTS) can be traced to a recommendation by one of our predecessor National Research Council panels on the decennial census. That panel suggested (National Research Council, 1985:Rec. 6.3) "that the Census Bureau keep machine-readable records on the follow-up history of individual households in the upcoming pretests, and for a sample of areas in the 1990 census, so that information for detailed analysis of the cost and error structures of conducting census follow-up operations on a sample basis will be available." Three years later, the idea had developed into a fuller proposal; in a brief report evaluating the projects for the REX (research, evaluation, and experimentation) program of the 1990 census, the panel commented on the idea and used the name that the project has since retained (National Research Council, 1988):

> The panel supports the concept of a master trace sample (MTS) that will facilitate a wide range of detailed studies of the quality of the 1990 census content. . . . The MTS will comprise a sample of census records that include not only the final values for each questionnaire item, but also the values for these items at each step in the processing, along with additional information such as whether the respondent mailed back a filled-in questionnaire or responded to telephone or personal follow-up. The MTS sample could well overlap other samples of interest, including the Current Population Survey (CPS), the Survey of Income and Program Participation (SIPP), the census reinterview sample, and others, and could have pertinent administrative records data appended to it. . . . We applaud the objectives of the MTS and support having as much of the file content as possible available in a public-use format.

The panel further noted that the sample "would greatly facilitate error analyses of the census" and would permit detailed examination of errors introduced in such processes as geocoding and imputation.

For various reasons—among them the overwhelming task of preparing for dual-systems estimation and subsequent coverage evaluation—the Bureau did not put the master trace sample idea into practice in 1990. But the concept of maintaining, pulling together, and analyzing detailed records of field operations took root. It was revisited, elaborated upon, and given much greater emphasis by the Panel on Alternative Census Methodologies as the 2000 census drew very close (National Research Council, 1999:93):

> The panel strongly supports a renewal and modest expansion of the suggestion by the Panel on Decennial Census Methodology of 10 years ago . . . for the collection of a master trace sample. With the various innovations in the 2000 census such as the possibility of sampling for nonresponse follow-up and alternative methods for enumeration (e.g., "Be Counted" forms), it would be very useful if the planned data management system could collect a trace sample in, say, 100 census tracts around the country. (Sampling tracts would facilitate study of the effects at the block or interviewer level.) The trace sample would provide information as to what happened in all phases of data collection, which will be instrumental in guiding methodological advances to be used in 2010 and beyond. Specific variables that could be included in the trace sample collection are as follows:
>
> - where the address came from (original master address list, local update, casing check, etc.);
> - the type of questionnaire (long or short form), whether and when it was returned, whether it was the first or a replacement questionnaire (or both), whether respondent-friendly enumeration was (also) used, if the household was a nonrespondent and a member of nonresponse follow-up sample, then how many approaches for field enumeration were made, which mode was used, whether they were ultimately successful, whether data capture required proxy enumeration and, if so, what type of proxy enumeration, edit failures, and finally whether there

> were any data differences among duplicate responses for households or individuals; and
>
> - the identification number of the enumerator, to facilitate evaluation of interviewer effects.
>
> Of course, any of the above information that could easily be collected on a 100 percent basis should be.

National Research Council (1999) continues with additional suggestions for data sources for inclusion in the Master Trace Sample, including measures of interviewer quality, results of unduplication programs, and information from the then-planned Integrated Coverage Measurement (later replaced by the Accuracy and Coverage Evaluation, or ACE, Program). The panel formalized its thoughts in a recommendation (National Research Council, 1999:Rec. 5.1):

> The panel recommends that a trace sample be collected in roughly 100 tracts throughout the United States and saved for research purposes. The trace sample would collect detailed process data on individual enumerations. In addition, similar information on integrated coverage measurement should be collected, on a sample basis if needed. It would be very useful if information could be collected, again on a sample basis, to support complete analysis of the census costs model, all aspects of the amount of duplication and efforts to unduplicate, and information needed to support total error modeling of the 2000 census.

Picking up where earlier panels left off, our Panel on Research on Future Census Methods has considered the Master Trace Sample to be central to its charge. Indeed, the MTS was a major topic of our first interim report (National Research Council, 2000a:1–2):

> We believe that the master trace sample database has the potential to be the single most useful source of information for assessing alternative designs for the 2010 census. ... The current plans for the master trace sample database should be augmented so that data for all key steps in the process—starting with address assignment and ending with a final disposition for each case—are included in the master trace sample database.

We made a number of other suggestions to the Census Bureau relative to the construction of the database (National Research Council, 2000a:15–18):

- use a two-stage sample design;
- oversample ACE blocks, list/enumerate and update/leave households and households that are hard to enumerate;
- improve the quality of information on the number and dates of attempts at enumeration;
- set priorities for the retention of master trace sample input files;
- provide for the accessibility and availability of the databases;
- increase the resources for developing the database; and
- collect sufficient information to support a model of total census error.

In March 2003, the panel was briefed on Census Bureau plans to implement a Master Trace Sample based on the 2000 census, including the proposed contents of the database and its intended uses. At that briefing, the issue of potential research questions—about which the panel was already somewhat aware—was spelled out with greater specificity. We were told that a total of fifteen "requirements" had guided Master Trace Sample research and development and that ten supplementary research questions fell into the category of acceptance testing. The requirements were mostly stated in the form of questions about simple relationships between two variables of interest; for example,

> What is the correlation between the date of completion of NRFU cases and the rate of item nonresponse?

or

> What is the correlation between the history of address sources and the need for applying the Primary Selection Algorithm (PSA) because of multiple responses for the same address?

The supplementary questions, for the most part, involved similar bivariate relationships.

Developed as part of the formal program of 2000 census evaluations, the Master Trace Sample final report was issued on September 29, 2003 (Hill and Machowski, 2003). Consisting of only eight text pages, the report confirmed the structural requirements and related questions noted at the March briefing. According to the report, "the MTS database links micro-level data from various stages of the Census 2000 project such as address frame development, data collection, data capture, data processing, and enumeration contact records" (Hill and Machowski, 2003:4). These data are linked at the following levels:

- local census office (LCO),
- enumerator,
- housing unit,
- return (that is, census questionnaire),
- enumeration contact (that is, personal visit), and
- person.

Moreover, "the MTS database is intended to address a wide variety of research requests that link decennial census response, data collection, and processing information with enumeration characteristics" (Hill and Machowski, 2003:4). The database contains the following types of data:

- geography;
- census response data at various stages of processing;
- enumeration characteristics (related to operations and enumerators);
- record of contact information from the nonresponse follow-up (NRFU) and coverage improvement follow-up (CIFU) operations;
- data capture system evaluation information from a reconciled keyed-from-image data set;
- geocoding error results from one of the Census 2000 evaluations; and

- housing unit status (i.e., occupied/vacant/delete/unresolved) from NRFU, CIFU, and ACE.

Among its limitations, the Master Trace Sample report notes that the database does not have Census 2000 person or housing unit coverage data from ACE; it excludes special places and group quarters; it does not include "the various Local Update of Census Addresses (LUCA) files or the bulk of the MAF extract files used to update the DMAF" (Hill and Machowski, 2003:6).

Under the heading "Intended Uses/Targeted Users," Hill and Machowski (2003:2) note that there is great potential for research in the following areas:

- modeling to identify and measure associations and relationships;
- tracing items, such as population count, through census processes; and
- investigating how to develop improved trace databases in future censuses.

Especially worth noting, under the same heading, Hill and Machowski (2003:3) state that

> the MTS database is limited to internal Census Bureau use. Census Bureau researchers interested in pursuing studies that will help guide the planning of the 2010 short form census will develop research proposals for review and approval by senior staff as well as planning groups guiding 2010 Census research.

We are greatly pleased to learn that the prototype of a Master Trace Sample was implemented in the 2000 census. We commend the Bureau for taking seriously the recommendations of our predecessor panels on census methodology; database construction has required a substantial commitment of Bureau personnel and resources. However, based on the information that has been given to us, we have some serious concerns about the direction that the MTS appears to be taking. Our concerns are rooted in a perceived divergence between the panel's vision of the MTS and its use and that of the Bureau, as we understand its position. The differences in these views fall under the headings of research, access, and plans.

Because our definition of research implies free-ranging and diligent inquiry, we are unconvinced of the wisdom of building the MTS on a set of preidentified research questions. Each of the fifteen database developmental requirements is reasonable in its own right, but, when taken as a set, the resulting structure is too narrowly focused. It is difficult to see how the inevitable questions that follow from initial queries can be pursued using the resulting database. Clearly, some crucial issues cannot be investigated at all given the data source limitations noted earlier, such as the extent to which duplication problems in the 2000 census may be traced to group quarters, or the characteristics of cases where the ACE failed to recognize and correct for person duplication.

Unfortunately, we are in the position of not knowing whether the MTS can contribute to a satisfactory understanding of any truly substantial design issue. In our view, a relational database contains a set of variables and their measures and permits the user to answer queries based not only on simple bivariate relationships but also on a broad range of joint and conditional associations. A menu-driven rather than query-based approach to analysis seems to us to be antithetical to good research.

To promote use of the MTS and expand its usefulness, we recommend the following:

> ***Recommendation 8.4:*** **The Census Bureau should develop a list of studies important to 2010 census planning that can exploit the richness of the Master Trace Sample. These studies should be prioritized and then conducted as resources permit.**

> ***Recommendation 8.5:*** **The Master Trace Sample from the 2000 census should be expanded to include data from group quarters enumeration, the Accuracy and Coverage Evaluation, and the Local Update of Census Addresses Program.**

With regard to MTS access, we are greatly troubled by the progression from the 1988 National Research Council panel's support for making as much of the file content as possible available in a public-use format, to the Census Bureau's March 2003 briefing document that allowed "internal research with indirect research opportunities for external customers such as the National Academy of Sciences," and finally to the final report's statement that "the MTS database access is limited to internal Census Bureau use." We are sensitive to confidentiality considerations in this regard but if, as stated in our first interim report, the MTS has the potential to be the single most important source of information for assessing alternative designs for the 2010 census, a great deal of this potential is lost to the Bureau by restricting its use.

In the language of the final MTS report, Census Bureau researchers will seek "approval by senior staff as well as planning groups guiding 2010 Census research" to investigate "hypotheses that involve relationships of various Census 2000 operations or systems" (Hill and Machowski, 2003:3). We have not been made aware of any specific projects that are now being pursued in this way. Moreover, in the absence of concrete knowledge of the database capabilities, we are unable to propose relevant and feasible studies for Bureau personnel or prioritize important areas for research. In brief, the extent to which the MTS can be properly mined remains unclear to us. We have not had access to the MTS, but we hope that the Bureau will modify its stance on access to permit broader use of the MTS.

> ***Recommendation 8.6:*** **The Census Bureau should explore ways to allow the broader research community to perform analyses using the 2000 Master Trace Sample, subject to confidentiality limitations.**

Our final recommendation is as follows:

> ***Recommendation 8.7:*** **The Census Bureau should carry out its future development in this area of tracing all aspects of census operations with the ultimate aim of creating a Master Trace *System*, developing a capacity for real-time evaluation by linking census operational databases as currently done by the Mas-**

ter Trace Sample. Emerging 21st century technology should make it feasible to know almost instantaneously the status of various census activities and how they interact. Such a system should be seriously pursued by the Census Bureau, whether or not it can be attained by 2010 (or even by 2020).

CHAPTER 9

Census Tests

AN IMPORTANT PART OF PLANNING for the decennial census is testing—trying out new procedures and techniques in order to finalize the census design before the count begins. A regular feature of the census process since the 1940 census, the Census Bureau's program of intercensal tests has pursued several major directions (Bailar, 2000):

- major changes in census methodology (most notably the conversion to mailout/mailback as the dominant mode of census collection and the use of sampling);
- techniques to improve and to better measure census coverage;
- improved questionnaire wording and format;
- new technology; and
- improved census processing.

From all indications, the Census Bureau is not eager to repeat the experience of the 2000 census, in which the lateness in reaching a general census design limited the effectiveness of operational testing. Under the heading "Lessons Learned from Census 2000," Waite (2002) emphasized the importance of effective testing: "if we want to achieve our Census 2010 Goals, operational testing of design infrastructure ***must start early*** in the decade and continue through the Dress Rehearsal." In particular, the census dress rehearsal—typically held 2 years prior to census deployment—should properly be a comprehensive run-through of census machinery to fine-tune the final census design. However, in 1998, the dress rehearsal had to serve as a feasibility test for three quite different general designs, involving different levels of sampling techniques (see Section 2–B; National Research Council, 2001a).

As depicted in Table 2-1, milestones in the 2010 planning process include major census tests roughly every other year leading up to 2010. Of these, one is already complete—the 2003 National Census Test, described in Box 9.1—and the 2004 Census Test is currently being fielded (see Box 9.2). Only two major testing opportunities remain prior to 2010: the 2006 census test, which the Census Bureau has described as a systems test, and the 2008 dress rehearsal.

In this chapter, we discuss some of the basic constraints on census testing (Section 9–A). We then briefly describe our basic recommendation to the Census Bureau with regard to the shape of the remaining census tests—namely, that the 2006 test should be cast as a vital proof-of-concept test (9–B). In the last section, we outline several priorities for census testing in the remaining years prior to the 2010 census (9–C).

9–A CONSTRAINTS ON CENSUS TESTING

The testing program for a decennial census faces a number of constraints and difficulties, some of which are unique to the census context but most of which are commonly faced by businesses or agencies in developing products or systems. The completion of a test plan for the 2010 census must try to strike a balance between these competing constraints.

Of these constraints, perhaps the most pressing—and the most common—is the need to match test activities to available resources. In the development cycle of a product or system, testing can sometimes be seen as an end-of-process activity and something to be done with the resources—monetary and person-hours alike—that remain at the end of a project. Relevant to the census context, this has often been the case in the development of computer-assisted interviewing instruments by the Census Bureau and other organizations (National Research Council, 2003b). On this account, the usual pattern of testing between decennial censuses compares favorably with that of some other sectors, in that a regular set of test activities is set up throughout the development process and, as a result, testing is a more constant presence through development. Still, the Census Bureau faces the unique problem of constructing a rigorous testing protocol under available appropriated funds, and the peak years for census testing—at the middle, and not the end, of the decade—are historically lean for funding of decennial census activities.

Limited census test budgets can affect not only the range of design options to be tested but also the means by which the test is conducted. For example, the resources available to conduct a census test may be more critical than methodological considerations in determining the number of test sites (as with the 2004 test; see Box 9.2). Likewise, resources may be more crucial to determining sample sizes for tests rather than power analysis to determine the optimal sample sizes needed to measure effects to desired precision. The choice to conduct the module of the 2003 National Census Test on race and Hispanic origin question wording was likely one of convenience using an available test activity involving a nationally representative sample from 2000 census mailout/mailback areas. However, a more meaningful test of sensitivity to wording and format considerations on the race and ethnicity questions would likely involve more refined targeting of predominantly minority and Hispanic neighborhoods as well as field interviewing.

A second, and related, major constraint on census testing is that—with the limited opportunities over the course of a decade—census tests are often formulated as test censuses. That is, the major census tests often have an omnibus character, typically involving most parts of the basic census process (see Section 2–A). Even though the tests may not provide test site locations with a new population count, it is often entirely possible that such counts could be derived, given the completeness of the process embodied in the test. There are several good reasons for the omnibus nature of census tests. It gives the major census tests the advantage of the verisimilitude of a decennial census, providing a realistic environment in which to test changes to specific techniques and allowing detection of unintended con-

Box 9.1 2003 National Census Test

Between February and April 2003, the Census Bureau conducted a National Census Test (NCT) involving approximately 250,000 households drawn from areas enumerated by mailout/mailback methods in the 2000 census. The NCT was strictly a mailout test, and so did not involve field enumerators to perform nonresponse follow-up.

The 2003 test focused primarily on two issues:

- *Response mode and contact strategies:* Different experimental groups were offered the opportunity to reply by mail (traditional method), Internet, or interactive voice response (IVR, an automated telephone system). Groups also varied as to whether these response modes were offered as a choice or whether they were "pushed" (e.g., providing Internet directions but no actual paper questionnaire in the mailing). Finally, contact strategies including targeted replacement questionnaires and reminder postcards were also varied. This component of the test involved eight experimental groups, one with 20,000 households and the other seven with 10,000 households each.
- *Race and ethnicity (Hispanic origin) question wording:* Seven treatment groups of 20,000 households each received different variations on the wording and arrangement of questions on race and Hispanic origins. Experimental settings varied by whether "some other race" was offered as a choice in the categories for race, whether wording was slightly revised to ask respondents if they are Hispanic or if they are of Hispanic origin, and whether instructions directed respondents to answer both questions.

The test was rounded out by a control group of 20,000 households; this group's questionnaire included the race and Hispanic origin questions worded as they were in the 2000 census (unlike the 2000 census context, the control group households were eligible for a replacement questionnaire in nonresponse follow-up in the 2003 test).

The samples for all groups were stratified by response rate in the 2000 census, where the classification was a grouping into "high" and "low" response groups based on a selected cut-off. Martin et al. (2003:11) comment that the low-response strata "included areas with high proportions of Blacks and Hispanics and renter-occupied housing units" and further comments that addresses in low-response areas were oversampled. Still, it is unclear whether the sample design generated enough coverage in Hispanic communities to facilitate conclusive comparisons—that is, whether it reached enough of a cross-section of the populace and a sufficiently heterogeneous mix of Hispanic nationalities and origins to gauge sensitivity to very slight and subtle changes in question wording.

Box 9.1 (continued)

With regard to the response mode and contact strategy portion of the test, results reported by Treat et al. (2003) suggest that multiple response mode options may change the distribution of responses by mode—shifting some would-be mail responses to Internet, for instance. However, the addition of choices does not generally increase cooperation overall. The experience of the 2003 test suggests serious difficulties with the interactive voice response option; 17–22 percent of IVR attempts had to be transferred to a human agent when the system detected that the respondent was having difficulty progressing through the IVR questionnaire. Moreover, rates of item nonresponse were greater for IVR returns than for the (paper response) control group. Internet returns, by comparison, experienced higher item response rates than the control. As has been indicated in past research, reminder postcards and replacement questionnaires had a positive effect on response.

Martin et al. (2003) report that the race and Hispanic origin question segment of the test showed mixed results. Predictably, elimination of "some other race" as a response category reduced "some other race" responses considerably, by 17.6 percent (that is, Hispanic respondents apparently declined to write in a generic response like "Hispanic" or "other" if "some other race" was not a formal choice). The Bureau concluded that the 17.6 percent decline in generic race reporting "more than offset" the impact of a 6.4 percent increase in the estimated number of Hispanics declining to answer the race question altogether (Martin et al., 2003:15). Adding examples of ancestry groups (e.g., Salvadoran, Mexican, Japanese, Korean) appeared to boost the reporting of detailed origins among Hispanics, Asians, and Native Hawaiian or Other Pacific Islanders. Treatment groups for which instructions were revised, instructing respondents to answer both the race and Hispanic origin questions, produced the most puzzling results; levels of missing data on one or both questions increased, as did the percentage of reporting themselves as Native Hawaiian or Other Pacific Islanders (relative to the control group).

sequences that a change might introduce in other parts of the process (e.g., difficulties that a change in questionnaire format might cause downstream in data capture). Another benefit of test censuses as census tests is that they provide an opportunity to "keep the wheels greased"—that is, they are a check to see that the complete census machinery is still in working order. But the test census model also creates difficulties; being more elaborate and involved, these tests can take longer to process and evaluate, thus potentially slowing feedback to the overall census planning process and to the development of subsequent tests. Another difficulty is the basic one of confounding: the simultaneous testing

Box 9.2 2004 Census Field Test

At this writing, the 2004 Census Field Test is being administered in test sites in two states: Colquitt, Thomas, and Tift Counties, Georgia, and a portion of northwestern Queens County (Queens Borough), New York. [Lake County, Illinois, was originally designated a test site, but was dropped after the Bush administration proposed its budget for fiscal 2004.] Though field work will be done in each of the test sites, and, in some respects, the activity will almost seem to be a census in miniature, the Census Bureau is not promising or even offering participating sites a population count at the end of the test. The test is intended to include approximately 200,000 housing units.

The operational plan for the 2004 test suggests four major topics (U.S. Census Bureau, 2003a:4–5):

- *Portable computing devices:* The clear primary thrust of the 2004 test is to assess the use of portable computing devices, equipped with GPS receivers, in field work for nonresponse follow-up (NRFU). The test will involve a workflow for the PCDs that relaxes some hierarchical aspects of past NRFU operations; completed questionnaires will be transmitted directly from enumerators' PCDs to headquarters and new assignments directly from headquarters to individual PCDs, without filtering through regional or local census offices. [However, some aspects of case management and assignment will still be done using paper reports, channeled through local offices.] The test is also intended to include dual-language (English or Spanish) questionnaires on the PCDs.
- *Coverage improvement:* In addition to testing whether edits built into the CAPI instrument on the PCDs help reduce nonresponse and the need for imputation, the test should also include preliminary attempts at real-time unduplication based on name and housing unit matching (see Section 5–E).
- *Further testing of race and Hispanic origin question wording:* This continues work from the 2003 National Census Test (see Box 9.1), though it is not clear whether all or most of the wording options from the 2003 National Census Test will be used in the 2004 test or whether the possibilities have been reduced.
- *Special place/group quarters definition:* The test is to include a first implementation of revised definitions of group quarters.

Other topics that were originally intended for inclusion in the 2004 test have subsequently been dropped from the test plan; these include the mailing of a dual-language (English and Spanish) questionnaire to targeted households and—significantly, given our discussion in Chapter 3—targeted canvass methods to update the Master Address File.

Box 9.2 (continued)

Current plans call for 10 evaluation reports of the results from the 2004 test to be issued through 2005; tentative dates for initial draft reports range from February 8 through November 2, 2005, and dates for final reports range from March 31 through December 31, 2005. However, the operational plan states that "preliminary results from a sub-set of evaluations needed to inform plans for the 2006 Census Test will be available no later than December 31, 2004" (U.S. Census Bureau, 2003a:7).

of multiple design options may make it difficult or impossible to assess the impact of one particular design option.

A final basic constraint on census testing is dealing with the maturity of technologies, systems, and methodologies. As we discussed in Section 6–D, new technology is inherently difficult to manage precisely because it is new and evolving. In terms of testing, new technologies bring with them a fundamental dilemma: the continued maturing of the technology depends on the results and feedback from testing, but there is a natural reluctance to test until the technology is mature (or at least reasonably so). This basic dilemma is also apparent when testing systems or groupings of technologies, as in the census context; reluctance to test one design option may affect development of other design options, which are dependent on the first in order to fully proceed. In the specific context of the 2010 census, for example, the Census Bureau cannot wait until portable computing devices are fully mature or until the MAF/TIGER database structure is complete and in operation to begin testing those components, and it certainly cannot wait for the MAF/TIGER piece to be completed before PCD development even though the former is a key information input to the latter. [These examples are merely to illustrate a basic interdependence; we do not imply that the Bureau is waiting until completion for the testing of either of these elements.]

9–B THE 2006 CENSUS TEST AS A PROOF OF CONCEPT

The Census Bureau needs to make optimal use of the few major testing opportunities remaining before 2010 so that viable approaches are well understood before the 2010 census design is finalized, and it must do so while facing the considerable challenges described in the previous section.

In the panel's assessment, the combination of the schedule of major census tests and the desire for a pure dress rehearsal in 2008 puts an enormous burden on the 2006 census test. The panel firmly believes that the 2006 test should be viewed as a proof-of-concept test: it should follow the census from end to end, to the greatest extent possible, using all available systems. More importantly, it should be cast as the proving ground for any remaining experimental questions in order to make the 2008 test a truly preoperational rehearsal. Any major 2010 census innovations should be identified in some moderately complete form by 2005 so that a reasonable version can be included in the 2006 census test.

In emphasizing the importance of the 2006 test, we believe that it is also important to make two points clear. First, the Census Bureau's hope—shared by the panel—is that the 2008 activity is to a great extent a pure dress rehearsal. That said, it is important to remember that it is also a test; things will go particularly right or wrong in 2008, and adjustments made accordingly. We note this to make clear that the 2006 test is not a completely hard-and-fast deadline for the inclusion of new technologies and techniques in the 2010 census. Some innovations will not be able to be tested in 2006 and will have to be tested in 2008; for example, it is unclear whether all functions of a redesigned MAF/TIGER database structure will be ready for 2006. What we hope to forestall by recommending that 2006 be viewed as the proof of concept is what happened in the 2000 census cycle, in which major design considerations were left for the highly experimental 1998 dress rehearsal to resolve.

The second point that we wish to make clear in calling for a proof-of-concept test in 2006 is that this test should not be viewed as the *only* remaining opportunity for new and experimental techniques. When possible and as resources permit, the Census Bureau should make use of other opportunities to evaluate alternative components of the 2010 design. Such opportunities might include small-scale experiments and feasibility tests, use of focus groups or small-scale laboratory-based studies for issues such as questionnaire format and other matters involving interviewer-respondent interactions, further analysis of the data collected in conjunction with the conduct of the 2000 census, and simulation studies (also often informed by data collected from the 2000 census). In addition, as discussed in Chapter 6, the development and comparison of alternative logical infrastructure models—each reflecting different assumptions and major design features—can be an informative way to test census systems in the abstract.

9–C DESIGN OPTIONS THAT SHOULD BE CONSIDERED FOR TESTING IN 2006

In this section, we briefly review design options that should be considered for evaluation, either as part of the 2006 census test or through use of other test opportunities as discussed above. We believe that the Bureau should know enough about each design option to fully inform a decision about the make-up of the census design for use in both the 2008 dress rehearsal and then, ideally with only minor modifications, in the 2010 census.

9–C.1 Human Factors for Portable Computing Devices

A critical area of concern in much of the new technology proposed for use in the 2010 census is that of human factors. Primary examples of areas where human factors require attention are: (1) the ability of field staff to quickly and reliably learn to use the portable computing devices (PCDs), including for navigation from one assignment to the next, the conduct and reporting of interviews, data transmission of completed assignments, and receipt of new assignments; (2) the ability of field staff to use ALMI (laptop computer) or, possibly, the GPS-equipped PCDs to capture updated coordinates for TIGER; and (3) the respondent interface provided not only by paper questionnaires but also by the electronic questionnaires used in the Internet and (possibly) interactive voice response modes, especially foreign language submissions. Of these, the highest priority is that of the human factors relative to the use of portable computing devices; as we noted in Section 5–A, the ultimate success of the PCDs will rely crucially on their usability by a corps of enumerators with relatively little training and, likely, a wide range of familiarity with using such devices. Therefore, human factors and the capacity of enumerators to successfully use the PCDs in nonresponse follow-up (as well as for other field activities such as address canvassing and coverage improvement follow-up that may be identified for PCD use) should be tested in 2006, most likely involving small-scale feasibility tests.

9–C.2 Various Cost-Benefit Trade-offs

In addition to demonstrating the basic feasibility and effectiveness of particular design options, the 2006 test should be constructed to permit assessment of various important cost-benefit trade-offs, many of which we have described in this report. Some of these may also be amenable to small-scale tests and other research activities. Regardless of how the tests are performed, it is important to learn more about these trade-offs because initial judgments about them have been used to support proposed components in the plan for the 2010 census. Some of these trade-offs are as follows:

Use of PCDs for Follow-up Interviewing

In Section 5–A.2, we describe the primary argument that the Census Bureau has used to support the plans for use of PCDs in nonresponse follow-up work: namely, that the devices will substantially reduce the amount of paper involved in the census. With the reduction in paper, the Bureau has argued that the number and size of local census offices may be reduced and that significant increases in data capture efficiency will further reduce costs. As we also noted, the panel knows of no empirical evidence for these potential cost savings.

While the 2004 census test is intended to provide some information on this trade-off, it is not clear that it will be definitive in this regard. To a large degree, the 2004 census test appears to be an extended, large-scale feasibility test. Such a test will be important to gauge responses by enumerators and respondents alike to the use of the small devices, and worthwhile as a preliminary check on the feasibility of the direct connection between census headquarters and individual PCDs for transmittal of completed questionnaires and new enumerator assignments. However, the PCD workflow for the 2004 test is still paper-driven in some respects, with printed progress reports circulating between headquarters and the regional and local census offices. Coupled with the limitation of the test to two sites, it is thus unclear how much information will be gained about the potential for local census office reduction and how much the results might be generalized in order to justify the overall PCD cost-benefit trade-off.

To the extent that the 2004 census test is unable to definitively inform this trade-off, the 2006 test must be able to bridge the gap. The 2006 test should certainly reflect adjustments to the headquarters-PCD workflow identified as part of the 2004 test. The 2006 test should also draw from the 2004 experience in gauging how much time and resources may be required for training temporary enumerators on PCD usage. Further examination of the security concerns associated with transmitting data to and from the PCD, either by phone line (as in the 2004 test) or possibly by wireless communication, should also be done, with particular attention to cost-benefit considerations.

Use of the Internet for Census Delivery and Return

The Internet is being proposed both as a method to facilitate response to the 2010 census for those who use the computer for much of their correspondence and as a mechanism by which foreign language questionnaires may be requested and administered to those who require them. Getting a better sense of what share of the population might be amenable to Internet response is an important cost-benefit consideration, since it will affect the amount of paper to be processed at the data capture stage (and otherwise handled and stored) and will have the benefits of automated data capture and the capacity for in-process edits and consistency checks. The cost of providing an Internet response option will be relatively minor, though the costs of protective security measures to prevent breaches by hackers must also be considered. Another major cost that must be weighed regarding Internet response is the potential for increased duplicate enumerations.

The 2003 census test made a first step at addressing some of these concerns, but the 2006 test or other testing opportunities should be constructed to provide more definitive assessment. The effect of pushing for Internet response (that is, providing directions for Internet response but not an actual paper questionnaire to be returned by mail) should be measured; the 2003 test included an Internet choice as well as an option that pushed both the Internet and telephone-based interactive voice response systems, the latter of which encountered difficulties. Alternative ways of making the Internet response choice more prominent in the mailed materials should also be explored.

Use of Imputation to Replace Last-Resort Enumeration

As discussed elsewhere in this report, the process used in the 2000 census to treat initial item nonresponse differed from that used in the 1990 census: in 2000, the choice between use of imputation methods versus more intensive field follow-up work was decided in favor of imputation. In particular, imputation was used to treat responses that were said to be data-defined—i.e., they meet a minimum standard for completeness in data reporting, which in 2000 meant that one person on the form had reported at least two data items. It is important that the impact of reliance on imputation on resulting data quality be more fully assessed. While some information may be gained from further analysis of 2000 census operational data—for example, by studying the effects of imputation on the distribution of census variables—a fuller test of the trade-off between imputation and field work requires testing in a full-census environment like the 2006 test. A test of this trade-off in 2006 should involve direct comparison of the two procedures, assessment of the additional costs of the field work, and evaluation of the quality of the information collected or imputed by implementation of a reinterview survey.

9–C.3 Other Testing Considerations

Throughout this report, we have identified additional key topics that should be considered for testing in 2006. We briefly list them here, and refer to the appropriate sections in the text for additional discussion.

- *Coverage Evaluation Methodology (Section 7–A):* Because assessment of the completeness of the census count is essential, the Census Bureau should develop candidate methods for coverage evaluation of the 2010 census, test them in the 2006 census tests, and test final methodology in the 2008 dress rehearsal.
- *Housing Unit Listing and Block Canvass Methodology (Sections 5–C.1 and 3–E.6):* The 2006 test should be a forum to test revised procedures for the listing and coding of housing units on the MAF, with particular attention to the problem of effectively identifying units in small multiunit structures. In addition, the 2006 test should provide an opportunity to target MAF updating to specific (e.g., high-growth) areas, rather than the complete block canvass that currently seems to be the Bureau's choice.
- *Residence Rules (Section 5–B.3):* The 2006 census test should feature revised rules for residence; redesigned questionnaires with clearer definitions of residence rules for respondents should be developed and tested by cognitive researchers prior to 2006.
- *Targeted Replacement Questionnaire Processing and Timing (Section 5–D.3):* Although the 2006 census will not approach the volume of a full census, the test should be used to assess the speed with which replacement questionnaires for nonresponding households can be printed and mailed. It should also provide an opportunity for gauging the appropriate time after questionnaire mailout to implement the replacement questionnaire mailing.
- *Routines for Unduplication (Section 5–E):* Having gone through some initial testing in the 2004 test, techniques for unduplication based on name and date-of-birth matching throughout the census response file should be refined and implemented in 2006. Testing should assess not only the accuracy of matching and the statistical models used to determine matches, but also the actual time needed for

searching and matching. Such information is crucial to determining whether "real-time unduplication" is possible in the 2010 census.

- *Use of Administrative Records in MAF Updating or Nonresponse Follow-up (Section 7–C.2):* If use of administrative records is considered for inclusion in the 2010 census process (as opposed to a major experiment), that use should be factored into the 2006 census test and included in the 2008 dress rehearsal.

In addition to the areas listed above, we commented in Section 5–C that enumeration methods for special hard-to-count populations—including gated communities, colonias, linguistically isolated households, and the homeless—should be the focus of research well before the end-of-decade crunch immediately prior to the census. To the extent possible, revised methods for these populations should be tested in 2006 rather than waiting for the 2008 dress rehearsal (or even later). Likewise, we recommended a comprehensive reappraisal and redefinition of the methods used for the group quarters population (Section 5–B.2). In particular, improved methods for developing the roster of group quarters should be developed in time for the 2006 census test, as should techniques for integrating or cross-checking the group quarters list with the MAF. The forms used to collect information for the group quarters population should also be reexamined to determine whether they are appropriate to part or all of the group quarters population.

9–C.4 Site Selection

As a final remark, site selection for the 2006 census test is extremely important. The Census Bureau typically selects a small number of counties for its test censuses to provide an effective test of its procedures. The counties are selected to represent urban and rural regions and to include various nonminority and minority groups. We urge the Census Bureau to select test sites that will provide an extreme and rigorous test of the various elements of the census design, so that the proof-of-concept test can best inform the reengineering of the 2010 census as a whole.

CHAPTER 10

Recommendations

FOR THE READER'S CONVENIENCE, we list below the specific recommendations that appear in the text of this report. The recommendations are grouped by chapter.

MODERNIZING GEOGRAPHIC RESOURCES

Recommendation 3.1: **The Census Bureau must devise a plan and develop effective procedures for updating and correcting the Master Address File (MAF). A complete and accurate Master Address File is critical not only to the success of the 2010 census but also to the effective implementation of the American Community Survey, the other household surveys conducted by the Census Bureau, and the 2008 dress rehearsal. Because the 2000 MAF was not simply discarded following the 2000 census (as occurred in censuses prior to 1990), the 2010 census will have as a base an address file of unprecedented completeness, but that does not obviate the need for continual updating, filtering, unduplicating, and cleaning of the MAF during the years leading to the 2010 census.**

The plan for a continually updated 2010 MAF must include, but not be limited to, the following:

1. A clear articulation of how the MAF/TIGER Enhancements Program and other Census Bureau activities will add missing

housing unit addresses, remove duplicate addresses, and generally correct the Master Address File, independent of benefits derived from being cross-referenced to an updated TIGER database;

2. **More effective definitions of housing units and methods to obtain accurate address listings for structures containing multiple housing units, as it is not sufficient to know only the address or geographic coordinates of the structure location;**
3. **Detail on the temporal sequencing and adequacy of address updates from the U.S. Postal Service's Delivery Sequence File, the Census Bureau's Community Address Updating System, and as-yet unspecified local partnership programs;**
4. **More effective means to define, list, and enumerate group quarters living arrangements, which should be done in coordination with the development and maintenance of the MAF; and**
5. **A detailed plan for Objective Five (quality metrics) of the MAF/TIGER Enhancements Program, including a program of evaluation and assessment of MAF coverage and input to the MAF/TIGER Redesign (Objective Two), so that the revised database structure includes appropriate address source codes and other useful variables for evaluation.**

Recommendation 3.2: **The Census Bureau should create and staff a position to oversee the development and maintenance of the MAF as a housing unit inventory, with a focus on improving methods to designate, list, and update units. This position should be responsible for development and implementation of plans drawn up consistent with Recommendation 3.1.**

Recommendation 3.3: **The Census Bureau should pursue more effective partnership and research collaboration with the U.S. Postal Service, including but not limited to further work on "undeliverable as addressed" items from the 2000 census, assessment of the address coverage quality of the Delivery Sequence File (DSF), and possibilities for more accurate translation of**

post office box listings and other DSF entries to street addresses and geographic coordinates.

Recommendation 3.4: **The Census Bureau should assess how critical the Community Address Updating System (CAUS) is to providing address updates in rural, non-city-style address areas. Such an assessment should include not only estimates of the number of addresses that could be provided and the workload that could be handled by CAUS/American Community Survey staff, but also empirical evidence on coverage gaps in the U.S. Postal Service's Delivery Sequence File by geographic area or type.**

Recommendation 3.5: **The Census Bureau should immediately develop and describe plans for partnerships with state, local, and tribal governments in collecting address list and geographic information. Such plans should include a focus on adding incentives for localities to contribute data to the census effort, making it easier for localities and the Bureau to exchange geographic information. Accordingly, plans for partnerships should include:**

- **clear articulation of realistic schedules for local input and review;**
- **definition and clear presentation of benchmark standards for local data to be submitted to the Bureau;**
- **mechanisms for providing effective feedback to local and tribal governments, detailing and justifying the Bureau's decisions to use or not use the information provided; and**
- **coordination of efforts across the Bureau so that calls for local and tribal entities to supply input to the Master Address File, TIGER, the Boundary and Annexation Survey, and other Bureau programs are not unduly redundant and burdensome.**

Recommendation 3.6: **The Census Bureau should evaluate the necessity of its plans to conduct a complete block canvass shortly before the 2010 census. Such justification must include analysis**

of extant census operational data and should include, but not be limited to, the following:

1. **arguments as to why selective targeting of areas for block canvass is either infeasible or inadequate, and as to how the costs of the complete block canvass square with the benefits; and**
2. **analysis of how a full block canvass fits into the Census Bureau's cost assumptions for the 2010 census.**

If plans proceed for a complete canvass, the Bureau should also consider how such a mass field deployment prior to 2010 could be used to achieve other improvements or efficiencies, such as the collection of GPS trace data as supplement to or as quality control for the TIGER realignment.

Recommendation 3.7: **The Census Bureau must:**

- **fully exploit the address source information in the MAF Extract in order to complete 2000 census evaluations, fill gaps in knowledge remaining from the 2000 census evaluations, and assess causes of duplicate and omitted housing units; and**
- **build the capability for timely and accurate address evaluation into the revised MAF/TIGER data architecture, including better ways to code address source histories and to format data sets for independent evaluation.**

AMERICAN COMMUNITY SURVEY

Recommendation 4.1: **The Census Bureau should continue research to understand the differences between and relative quality of ACS estimates and long-form estimates, with particular attention to measurement differences and error from nonresponse and imputation. The Bureau must work on ways to effectively communicate and articulate those findings to interested stakeholders, particularly potential end users of the data.**

Recommendation 4.2: **The Census Bureau must make ACS data available (protecting confidentiality) to analysts in the 31 ACS**

test sites to facilitate the comparison of ACS and census long-form estimates as a means of assessing the quality of ACS data as a replacement for census long-form data. Again, with appropriate safeguards, the Census Bureau should release ACS data to the broader research community for evaluation purposes.

Recommendation 4.3: **The Census Bureau must issue a guide for users of ACS data that details the statistical implications of the difference between point-in-time and moving average estimates for various uses.**

Recommendation 4.4: **As soon as possible, based on the 2006 proof-of-concept test, the Census Bureau should work with Congress and the administration to secure agreement on the overall design for the 2010 census and the American Community Survey (ACS). Extended delay in finalizing an overall design for the 2010 census—such as occurred in the preparation for the 2000 census—would unacceptably heighten the risk associated with the 2010 census. The role of the ACS is of particular concern; failure to secure commitment to the ACS as a replacement for the census long form would severely impair plans for a short-form-only census and undercut the ability to provide reliable small-area characteristics data by 2010.**

The Census Bureau should identify the costs and benefits of various approaches to collecting characteristics information if support for the full ACS is not forthcoming. These costs and benefits should be presented for review so that decisions on the ACS and its alternatives can be fully informed.

Recommendation 4.5: **The Census Bureau should identify the costs and benefits of various approaches to collecting characteristics information should funding for the full ACS not be forthcoming. These costs and benefits should be presented for review so that decisions on the ACS and its alternatives can be fully informed.**

ENUMERATION AND DATA-PROCESSING METHODS

Recommendation 5.1: **The Census Bureau should develop and perform a rigorous test of its plans for use of portable computing devices, and this test should compare the performance and outcomes of data collection using:**

- **devices of the current (Pocket PC) class being developed for use in the 2004 census test;**
- **high-end devices (e.g., tablet computers) of classes that are very likely to be available at reasonable cost by the time of procurement for 2010; and**
- **traditional paper instruments.**

Such a test is intended to provide fuller information about the costs and benefits of portable computing devices, using paper as a point of comparison. The test should also provide the opportunity to review specifications and requirements for the PCDs, using devices of the caliber likely to be available by 2010.

Recommendation 5.2: **By the end of 2004, the Census Bureau should complete requirements design for its portable computing devices, building from the results of the 2004 census test and in anticipation of the 2006 proof-of-concept test. The requirements and specifications for portable computing devices must include full integration with the census system architecture and should include suitability for other, related Census Bureau applications. The Bureau's requirements design for PCDs must devote particular attention to the human factors underlying use of the devices.**

Recommendation 5.3: **The Census Bureau must develop a complete engineering and testing plan for the software components of the portable computing devices, with particular attention to the computer-assisted personal interviewing interface, data capture systems, and communication/synchronization capabilities (including assignment of enumerator workload).**

Recommendation 5.4: **The Census Bureau's techniques for enumerating the population in special places and group quarters must be completely evaluated and redesigned for the 2010 census. This effort must include (but not be limited to):**

- **clear definitions of group quarters;**
- **redesign of questionnaire and data content as appropriate, including a provision for handling data items that might best be provided by group quarters administrators rather than individual residents;**
- **collection of information, including additional addresses, that will be needed to facilitate unduplication of all census records;**
- **improvement of the address listing processes for group quarters, including coordination with the development of the Master Address File; and**
- **specification of enumeration and coverage evaluation plans for group quarters.**

Recommendation 5.5: **The Census Bureau's development of tailored enumeration methods for special populations—including irregular urban areas, colonias, gated communities, and rural areas—must begin early, and not be put off for development late in the census planning cycle.**

Recommendation 5.6: **The Census Bureau must quickly determine ways to implement a second questionnaire mailing to non-responding households in the 2010 census, in order to improve mail response rates. Such determination should be done in a cost-effective manner that minimizes duplicate enumerations, but must be made early enough to avoid the late problems that precluded such a mailing in the 2000 census.**

Recommendation 5.7: **The Census Bureau must develop comprehensive plans for unduplication in the 2010 census, in terms of both housing units and person records. Housing unit unduplication research and efforts should be conducted consistent**

with objectives outlined in the panel's recommendations related to the Master Address File. Person-level unduplication efforts should focus on improvements to the methodology developed for the 2000 Accuracy and Coverage Evaluation Program, including national-level matching of records by person name. It is essential that changes in unduplication methodology be tested and evaluated using extant data from the 2000 census and that unduplication methods be factored into the 2006 proof-of-concept test and 2008 dress rehearsal.

Recommendation 5.8: **The Census Bureau must pursue research on the trade-off in costs and accuracy between field (enumerator) work and imputation routines for missing data. Such research should be included in the 2006 proof-of-concept test, and census imputation routines should be evaluated and redefined prior to the 2008 dress rehearsal. As appropriate, the American Community Survey research effort should also address the trade-off between imputation and field work.**

Recommendation 5.9: **The Census Bureau should conduct research into the effects of imputation on the distributions of characteristics, and routines for imputation of specific data items should be completely evaluated and revised as appropriate for use in the American Community Survey.**

TECHNICAL INFRASTRUCTURE AND BUSINESS PROCESS

Recommendation 6.1: **In order to achieve the full benefit of architecture modeling, the highest management levels of the Census Bureau should commit to the design and testing of a redesigned logical architecture, so that the most promising model can facilitate the implementation of an efficient technical infrastructure for the 2010 census.**

Recommendation 6.2: **To ensure the successful integration of new technologies and techniques in the census process, the Census Bureau should create and staff the position of system ar-**

chitect for the decennial census. The selected candidate should have expertise in modeling business processes, designing large-scale systems, and conducting reengineering activities. The system architect must be given the authority to work with and coordinate efforts among the organizational divisions within the Census Bureau and should serve as a champion of the importance of architecture reengineering at the highest levels of management within the Bureau.

The Census Bureau should also consider designating a subsystem architect for portable computing devices and related field systems, as has already been done for the MAF/TIGER redesign. The efforts of the MAF/TIGER redesign and PCD subsystem architects should be coordinated in partnership with the system architect for the decennial census.

Recommendation 6.3: As part of the MAF/TIGER redesign, the Census Bureau should consider ways to make its application code for mapping, geocoding, digital exchange, map editing, and other functions openly available in order to facilitate continued ties to and improvement in geographic information systems software applications and to tap the feedback of the broader computer science/software development community.

Recommendation 6.4: The Census Bureau should generally improve its software engineering processes and should pursue its goal of raising its Capability Maturity Model score in software development. In particular, the Bureau should focus on available tools and techniques in rigorously developing and tracking software requirements and specifications. In beginning the task of improving its software practices, though, the Census Bureau must recognize that the effort is a difficult one, requiring high-level commitment in the same manner as architecture reengineering.

Recommendation 6.5: The Census Bureau should evaluate and improve its protocols for hardware and software testing, drawing on expertise from the computer science and software development communities. Rigorous hardware and software testing should be factored into census operational schedules, in addi-

tion to the field testing performed in the 2006 proof-of-concept test, the 2008 dress rehearsal, or such other formal census tests as may arise.

COVERAGE MEASUREMENT

Recommendation 7.1: **The Census Bureau should continue to pursue methods of improving demographic analysis estimates, working in concert with other statistical agencies that use and provide data inputs to the postcensal population estimates. Work should focus especially on improving estimates of net immigration. Attention should also be paid to quantifying and reporting uncertainty in demographic estimates. Updated assessments of the assumptions underlying demographic analysis estimates, including the completeness of birth registration, should also be considered.**

EVALUATIONS

Recommendation 8.1: **The Census Bureau should materially strengthen the evaluation component of the 2010 census, including the ongoing testing program for 2010. Plans for census evaluation studies should include clear articulation of each study's relevance to overall census goals and objectives; connections between research findings and operational decisions should be made clear. The evaluation studies must be less focused on documentation and accounting of processes and more on exploratory and confirmatory research while still clearly documenting data quality. To this end, the 2010 census evaluation program should:**

1. **identify important areas for evaluations (in terms of both 2010 census operations and 2020 census planning) to meet the needs of users and census planners and set evaluation priorities accordingly;**
2. **design and document data collection and processing systems so that information can be readily extracted to support timely, useful evaluation studies;**

3. focus on analysis, including use of graphical and other exploratory data analysis tools to identify patterns (e.g., mail return rates, imputation rates) for geographic areas and population groups that may suggest reasons for variations in data quality and ways to improve quality (such tools could also be useful in managing census operations);
4. consider ways to incorporate real-time evaluation during the conduct of the census;
5. give priority to development of technical staff resources for research, testing, and evaluation; and
6. share preliminary analyses with outside researchers for critical assessment and feedback.

Recommendation 8.2: A major focus of the Census Bureau's ongoing research and evaluation program must be opportunities for targeting and efficiency—tailoring approaches to key population groups and areas rather than pursuing a "one-size-fits-all" approach.

Recommendation 8.3: The Census Bureau must mine and fully exploit data resources currently available in order to build a research base for the 2010 census and to further evaluate the 2000 census. These resources include:

- **microdata from the 2000 Accuracy and Coverage Evaluation and its related Person Duplication Studies;**
- **extracts from the Master Address File;**
- **the Local Census Office Profile dataset;[1]**
- **a match of census records and the March 2000 Current Population Survey; and**
- **the Master Trace Sample.**

[1]Created as part of the 2000 census evaluations program, the Local Census Office Profile (Imel, 2003) is an extract from a merged data set drawing from, among other sources, the Decennial Master Address File and the Hundred Percent Unedited and Edited Files (the person-level files of census reports that are tabulated to produce census results).

Recommendation 8.4: **The Census Bureau should develop a list of studies important to 2010 census planning that can exploit the richness of the Master Trace Sample. These studies should be prioritized and then conducted as resources permit.**

Recommendation 8.5: **The Master Trace Sample from the 2000 census should be expanded to include data from group quarters enumeration, the Accuracy and Coverage Evaluation, and the Local Update of Census Addresses Program.**

Recommendation 8.6: **The Census Bureau should explore ways to allow the broader research community to perform analyses using the 2000 Master Trace Sample, subject to confidentiality limitations.**

Recommendation 8.7: **The Census Bureau should carry out its future development in this area of tracing all aspects of census operations with the ultimate aim of creating a Master Trace *System*, developing a capacity for real-time evaluation by linking census operational databases as currently done by the Master Trace Sample. Emerging 21st century technology should make it feasible to know almost instantaneously the status of various census activities and how they interact. Such a system should be seriously pursued by the Census Bureau, whether or not it can be attained by 2010 (or even by 2020).**

Bibliography

Alexander, C. H. (1993). *A Continuous Measurement Alternative for the U.S. Census*. Report CM-10. Washington, DC: Bureau of the Census, U.S. Department of Commerce.

Alexander, C. H. (2002). A discussion of the quality of estimates from the American Community Survey for small population groups. Draft prepared for discussion at the fall 2002 Census Advisory Committee Meetings, U.S. Census Bureau, Washington, DC.

Angueira, T. (2003a, June 3). *Planning for the 2010 Decennial Census ("Plan for the Plan")*. 2010 Census Planning Memoranda Series, Number 13. Washington, DC: U.S. Census Bureau.

Angueira, T. (2003b, June 3). *Reengineering the Decennial Census: The Baseline Design for 2010*. 2010 Census Planning Memoranda Series, Number 14. Washington, DC: U.S. Census Bureau.

Bailar, B. A. (2000). Census testing. In M. J. Anderson (Ed.), *Encyclopedia of the U.S. Census*, pp. 62–66. Washington, DC: CQ Press.

Baird, J. (2003, October 2). Utah weighs a 3rd census fight. *Salt Lake Tribune*. Available: http://www.sltrib.com/2003/Oct/10022003/utah/97854.asp [12/15/03].

Bauder, M. and D. Judson (2003). *Administrative Records Experiment in 2000 (AREX 2000): Household Level Analysis*. Census 2000 Experiment Report. Washington, DC: U.S. Census Bureau.

Bennett, C. E. and D. H. Griffin (2002). Race and Hispanic origin data: A comparison of results from the Census 2000 Supplementary

Survey and Census 2000. Paper prepared for the August 2002 Joint Statistical Meetings, New York City, NY. U.S. Census Bureau, Washington, DC.

Berning, M. A. (2003). *Administrative Records Experiment in 2000 (AREX 2000): Request for Physical Address Evaluation*. Census 2000 Experiment Report. Washington, DC: U.S. Census Bureau.

Berning, M. A. and R. H. Cook (2003). *Administrative Records Experiment in 2000 (AREX 2000): Process Evaluation*. Census 2000 Experiment Report. Washington, DC: U.S. Census Bureau.

Burcham, J. L. (2002, April 5). *Block Canvassing Operation*. Census 2000 Evaluation F.5. Washington, DC: U.S. Census Bureau.

Bureau of the Census (1964). *Evaluation and Research Program of the U.S. Censuses of Population and Housing, 1960: Accuracy of Data on Population Characteristics as Measured by CPS-Census Match*. ER 60–5. Washington, DC: U.S. Department of Commerce.

Bureau of the Census (1970). *Evaluation and Research Program of the U.S. Censuses of Population and Housing, 1960: Record Check Study of Accuracy of Income Reporting*. ER 60–8. Washington, DC: U.S. Department of Commerce.

Bureau of the Census (1975a). *1970 Census of Population and Housing Evaluation and Research Program: Accuracy of Data for Selected Housing Characteristics as Measured by Reinterviews*. PHC(E)-10. Washington, DC: U.S. Department of Commerce.

Bureau of the Census (1975b). *1970 Census of Population and Housing Evaluation and Research Program: Accuracy of Data for Selected Population Characteristics as Measured by the 1970 CPS-Census Match*. PHC(E)-11. Washington, DC: U.S. Department of Commerce.

Carbaugh, L. W. and R. W. Marx (1990). The TIGER system: A Census Bureau innovation serving data analysts. *Government Information Quarterly* 7, 285–306.

Caspar, R. A. (2003). *Synthesis of Results from the Response Mode and Incentive Experiment*. Census 2000 Testing, Experimentation, and Evaluation Program Report. Washington, DC: U.S. Census Bureau.

Centech Group, Inc. (2002a). *Census 2000 Operational Flow Baseline Report—Volume 1: Executive Package. Version 1.2 (Final)*. Arlington, VA: Centech Group, Inc. Prepared for the U.S. Census Bureau.

Centech Group, Inc. (2002b). *Census 2000 Operational Flow Baseline Report—Volume 2: Detailed Package. Version 1.3 (Final).* Arlington, VA: Centech Group, Inc. Prepared for the U.S. Census Bureau.

Centech Group, Inc. (2002c). *Logical Architecture Re-engineering Exercise. Version 1.1 (Draft).* Arlington, VA: Centech Group, Inc. Prepared for the U.S. Census Bureau.

Citro, C. F. (2000). Population estimates and projections. In M. J. Anderson (Ed.), *Encyclopedia of the U.S. Census*, pp. 300–303. Washington, DC: CQ Press. Assisted by Meyer Zitter.

Decennial Management Division (2003, September 26). *2010 Census Risk Management Plan, Release 2.5.* Washington, DC: U.S. Census Bureau.

Ericksen, E. P. and J. B. Kadane (1983). Using administrative lists to estimate census omissions: An example. In *Proceedings of the Survey Research Methods Section*, pp. 361–366. Alexandria, VA: American Statistical Association.

Executive Steering Committee for A.C.E. Policy (2001a). *Report of the Executive Steering Committee for Accuracy and Coverage Evaluation Policy on Adjustment for Non-Redistricting Uses.* Washington, DC: U.S. Census Bureau.

Executive Steering Committee for A.C.E. Policy (2001b). *Report: Recommendation Concerning the Methodology to be Used in Producing Tabulations of Population Reported to States and Localities Pursuant to 13 U.S.C. 141(c).* Washington, DC: U.S. Census Bureau.

Fay, R. E. (2001, October 26). *ESCAP II: Evidence of Additional Erroneous Enumerations from the Person Duplication Study.* Executive Steering Committee for A.C.E. Policy II Supporting Report 9. Washington, DC: U.S. Census Bureau.

Franz, L. M. (2002). 2010 reengineering. Presentation to the Panel on Research on Future Census Methods, September 13, 2002, Washington, DC. U.S. Census Bureau.

Gilbert, C. (2000, March 30). Politics on the menu: Visiting Wisconsin, Bush says he sympathizes with census critics. *Milwaukee Journal Sentinel*. Available: http://www.jsonline.com/news/metro/mar00/george31033000.asp [12/01/03].

Gosselin, J. F. (1980). Reverse record check: Tracing people in Canada. *Survey Methodology 6*, 84–113.

Griffin, D. H. and S. M. Obenski (2001, September 28). *A Demonstration of the Operational Feasibility of the American Community Survey*. Census 2000 Testing, Experimentation, and Evaluation Program Report. Washington, DC: U.S. Census Bureau.

Hahn, R. A. (1992). The state of federal statistics on racial and ethnic groups. *Journal of the American Medical Association 267*, 268–271.

Hahn, R. A., J. Mulinara, and S. M. Teutsch (1992). Inconsistencies in coding of race and ethnicity between birth and death in U.S. infants. *Journal of the American Medical Association 267*, 259–263.

Hauser, P. M. (1942). Proposed annual sample census of population. *Journal of the American Statistical Association 37*(217), 81–88.

Heimovitz, H. K. (2003). *Administrative Records Experiment in 2000 (AREX 2000): Outcomes Evaluation*. Census 2000 Experiment Report. Washington, DC: U.S. Census Bureau.

Hill, J. M. and J. D. Machowski (2003, September 29). *Master Trace Sample*. Census 2000 Evaluation B.6. Washington, DC: U.S. Census Bureau.

Hirschfeld, D. (2000). Address list development. In M. J. Anderson (Ed.), *Encyclopedia of the U.S. Census*, pp. 6–13. Washington, DC: CQ Press.

Hogan, H. (1983). The Forward Trace Study: Its purpose and design. In *Proceedings of the Survey Research Methods Section*, pp. 168–172. Alexandria, VA: American Statistical Association.

Imel, J. D. (2003, September 30). *Local Census Office Profile for Census 2000*. Census 2000 Evaluation H.9. Washington, DC: U.S. Census Bureau.

Jackson, A. A. (2003, September). MAF/TIGER Enhancements Program Update, Objectives 1–4. Presentation to the Panel on Research on Future Census Methods, September 11, 2003, Washington, DC. U.S. Census Bureau.

Jonas, K. (2002, November 7). *Group Quarters Enumeration*. Census 2000 Evaluation E.5. Washington, DC: U.S. Census Bureau.

Jonas, K. (2003, August 6). *Group Quarters Enumeration*. Census 2000 Evaluation E.5, Revision 1. Washington, DC: U.S. Census Bureau.

Judson, D. H. and B. Bye (2003, October 21). *Synthesis of Results from the Administrative Records Experiment in 2000 (AREX 2000).* Washington, DC: U.S. Census Bureau.

Kish, L. (1981). *Population Counts from Cumulated Samples.* Pp. 5–50 in Congressional Research Service, *Using Cumulated Rolling Samples to Integrate Census and Survey Operations of the Census Bureau.* Prepared for the Subcommittee on Census and Population, Committee on Post Office and Civil Service, U.S. House of Representatives. Washington, DC: U.S. Government Printing Office.

Kish, L. (1990). Rolling samples and censuses. *Survey Methodology 16,* 63–71.

LaMacchia, R. (2003). MAF/TIGER update. Presentation to the Mapping Sciences Committee, National Research Council, October 23, 2003, Washington, DC. U.S. Census Bureau.

Liadis, J. S. (2000). *GPS TIGER Accuracy Analysis Tools (GTAAT) Evaluation and Test Results.* Washington, DC: U.S. Census Bureau.

Logan, J. R. (2003). How race counts for Hispanic Americans. Research report, Lewis Mumford Center for Comparative Urban and Regional Research, State University of New York, Albany. Available: http://mumford1.dyndns.org/cen2000/BlackLatinoReport/BlackLatinoReport.%pdf [2/16/04].

Love, S. (2002). Methodological issues in local area application of the American Community Survey. Presentation prepared for the August 2002 Joint Statistical Meetings, New York City, NY. U.S. Census Bureau, Washington, DC.

Lowenthal, T. A. (2003a). American Community Survey start would be delayed further under president's budget proposal for 2004. Census News Brief posted to http://www.census2000.org, February 10.

Lowenthal, T. A. (2003b). President Bush sends FY04 budget to Congress; Census Bureau would scale back test plans under proposed funding levels. Census News Brief posted to http://www.census2000.org, February 4.

Martin, E., D. Sheppard, M. Bentley, and C. Bennett (2003). *Results of 2003 National Census Test of Race and Hispanic Questions.* Unpublished report, dated October 1. Washington, DC: U.S. Census Bureau.

Martin, P. and E. Midgley (2003, June). *Immigration: Shaping and Reshaping America*. Washington, DC: Population Reference Bureau. Issued as *Population Bulletin*, Vol. 58, No. 2.

Marx, R. W. (1986). The TIGER system: Automating the geographic structure of the United States census. *Government Publications Review 13*, 68–78.

Nash, F. F. (2000). *Overview of the Duplicate Housing Unit Operations*. Washington, DC: U.S. Census Bureau.

National Research Council (1985). *The Bicentennial Census: New Directions for Methodology in 1990*. Panel on Decennial Census Methodology, Constance F. Citro and Michael L. Cohen, eds., Committee on National Statistics. Washington, DC: National Academy Press.

National Research Council (1988). *Priorities for the 1990 Census: Research, Evaluation, and Experimentation (REX) Program*. Panel on Decennial Census Methodology, Constance F. Citro and Michael L. Cohen, eds., Committee on National Statistics. Washington, DC: National Academy Press.

National Research Council (1994). *Counting People in the Information Age*. Panel to Evaluate Alternative Census Methods, Duane L. Steffey and Norman M. Bradburn, eds., Committee on National Statistics. Washington, DC: National Academy Press.

National Research Council (1995). *Modernizing the U.S. Census*. Panel on Census Requirements in the Year 2000 and Beyond, Barry Edmonston and Charles Schultze, eds., Committee on National Statistics. Washington, DC: National Academy Press.

National Research Council (1996). *Continued Review of the Tax Systems Modernization of the Internal Revenue Service: Final Report*. Committee on Continued Review of the Tax Systems Modernization of the Internal Revenue Service, Computer Science and Telecommunications Board. Washington, DC: National Academy Press.

National Research Council (1999). *Measuring a Changing Nation: Modern Methods for the 2000 Census*. Panel on Alternative Census Methodologies, Michael L. Cohen, Andrew A. White, and Keith F. Rust, eds., Committee on National Statistics. Washington, DC: National Academy Press.

National Research Council (2000a). *Designing the 2010 Census: First Interim Report.* Panel on Research on Future Census Methods, Michael L. Cohen and Benjamin F. King, eds., Committee on National Statistics. Washington, DC: National Academy Press.

National Research Council (2000b). *Small-Area Income and Poverty Estimates: Priorities for 2000 and Beyond.* Panel on Estimates of Poverty for Small Geographic Areas, Constance F. Citro and Graham Kalton, eds., Committee on National Statistics. Washington, DC: National Academy Press.

National Research Council (2001a). *The 2000 Census: Interim Assessment.* Panel to Review the 2000 Census, Constance F. Citro, Daniel L. Cork, and Janet L. Norwood, eds., Committee on National Statistics. Washington, DC: National Academy Press.

National Research Council (2001b). *The American Community Survey: Report of a Workshop.* Committee on National Statistics. Washington, DC: National Academy Press.

National Research Council (2001c). Letter from Benjamin F. King to William G. Barron, acting director, U.S. Census Bureau. Dated February 15. Panel on Research on Future Census Methods, Committee on National Statistics. Washington, DC: National Academy Press.

National Research Council (2003a). *Planning the 2010 Census: Second Interim Report.* Panel on Research on Future Census Methods, Daniel L. Cork, Michael L. Cohen, and Benjamin F. King, eds., Committee on National Statistics. Washington, DC: The National Academies Press.

National Research Council (2003b). *Survey Automation: Report and Workshop Proceedings.* Oversight Committee for the Workshop on Survey Automation, Daniel L. Cork, Michael L. Cohen, Robert Groves, and William Kalsbeek, eds., Committee on National Statistics. Washington, DC: The National Academies Press.

National Research Council (2004). *The 2000 Census: Counting Under Adversity.* Panel to Review the 2000 Census, Constance F. Citro, Daniel L. Cork, and Janet L. Norwood, eds., Committee on National Statistics. Washington, DC: The National Academies Press.

Norman, J. (2000, March 7). Stumbling beginning: Initial census mailing leaves some scratching their heads. *Milwaukee Journal Sentinel.* Available: http://www.jsonline.com/news/metro/mar00/census08030700a.asp [12/01/03].

O'Grady, K. (2000). *A DOQ Test Project: Collecting Data to Improve TIGER*. Washington, DC: U.S. Census Bureau.

O'Grady, K. and L. Godwin (2000). *The Positional Accuracy of MAF/TIGER*. Washington, DC: U.S. Census Bureau.

O'Hara, C. and D. Caterinicchia (2001, August 20). Census' hunt for accuracy. *Federal Computer Week*, 10.

Owens, K. L. (2002, May 14). *Evaluation of the Local Update of Census Addresses 99 (LUCA 99)*. Census 2000 Evaluation F.6. Washington, DC: U.S. Census Bureau.

Owens, K. L. (2003, April 16). *Evaluation of the Local Update of Census Addresses 98 (LUCA 98)*. Census 2000 Evaluation F.3. Washington, DC: U.S. Census Bureau.

Passel, J. S. and W. Zimmerman (2001, May). *Are Immigrants Leaving California? Settlement Patterns of Immigrants in the Late 1990s*. Washington, DC: Urban Institute.

Prewitt, K. (2000, March 8). Prepared statement. Presented at hearing of the Subcommittee on the Census, Committee on Government Reform, U.S. House of Representatives. Available: http://www.census.gov/dmd/www/3-8-00.html [12/15/03].

Robinson, J. G. (2001). *ESCAP II: Demographic Analysis Results*. Executive Steering Committee for A.C.E. Policy II Supporting Report 1. Washington, DC: U.S. Census Bureau.

Rosenson, P. (2001). *Automated Updating of TIGER from Locally Developed Geospatial Data Bases*. Geography Division Report (unpublished). Washington, DC: U.S. Census Bureau.

Salvo, J. J. and A. P. Lobo (2002). The American Community Survey: Quality of response by mode of data collection in the Bronx test site. Paper prepared for the August 2002 Joint Statistical Meetings, New York City, NY. Department of City Planning, New York, NY.

Salvo, J. J., A. P. Lobo, and S. P. Love (2003). Evaluating continuous measurement: Data quality in the Bronx test site of the American Community Survey. *Journal of Economic and Social Measurement 28*, 263–277.

Singh, R. P. and W. Bell (2003). 2010 census coverage measurement goals and objectives. Presentation to the NRC Panel on Research on Future Census Methods, September 11, 2003, Washington, DC. U.S. Census Bureau.

Thibaudeau, Y. (1998). *Model Explicit Item Imputation for Demographic Categories for Census 2000.* Statistical Research Division Research Report RR-99-02. Washington, DC: U.S. Census Bureau.

Thompson, J. H. (2000). Organization and administration of the census. In M. J. Anderson (Ed.), *Encyclopedia of the U.S. Census*, pp. 295–299. Washington, DC: CQ Press.

Titan Corporation (2003, July 22). *Census 2000 Data Capture.* Census 2000 Testing, Experimentation, and Evaluation Program, Topic Report 3. Washington, DC: U.S. Census Bureau.

Titan Systems Corporation (2002, June 6). *Laptop Computers for Accuracy and Coverage Evaluation System Requirements Study.* Census 2000 Evaluation R.2.b. Washington, DC: U.S. Census Bureau.

Treat, J. B., S. Brady, J. A. Bouffard, and C. Stapleton (2003). *2003 National Census Test: The Impact of Alternative Modes and Contact Strategies on Self-Response.* Unpublished report. Washington, DC: U.S. Census Bureau.

U.S. Census Bureau (2000). *System Architecture, Version 2.0.* Washington, DC: U.S. Census Bureau.

U.S. Census Bureau (2001a, June). *Potential Life Cycle Savings for the 2010 Census.* Washington, DC: U.S. Census Bureau.

U.S. Census Bureau (2001b, February). *TIGER Documentation v. 7.0.* Washington, DC: Geography Division, U.S. Census Bureau.

U.S. Census Bureau (2002a, May). *Census 2000 Testing, Experimentation, and Evaluation Program.* May 2002 Update. Washington, DC: U.S. Census Bureau.

U.S. Census Bureau (2002b). *Meeting 21st Century Demographic Data Needs—Implementing the American Community Survey: May 2002; Report 2: Demonstrating Survey Quality.* Washington, DC: U.S. Census Bureau.

U.S. Census Bureau (2003a, September 29). *2004 Census Test Operational Plan.* Washington, DC: U.S. Census Bureau.

U.S. Census Bureau (2003b, October 3). *2010 Census Business Architecture, Version 1.0.* Washington, DC: U.S. Census Bureau.

U.S. Census Bureau (2003c). *American Community Survey Operations Plan, Release 1: March 2003.* Washington, DC: U.S. Census Bureau.

U.S. Census Bureau (2003d, February). *Census 2000 Testing, Experimentation, and Evaluation Program Summary Documentation: Program Modifications Since May 2002*. Washington, DC: U.S. Census Bureau.

U.S. Census Bureau (2003e, October). MAF/TIGER Enhancements Program Update, Objectives 1–4. Update of the October 21 Presentation to the NRC Panel on Research on Future Census Methods, September 11, 2003, Washington, DC (no attribution).

U.S. Census Bureau (2003f). *Meeting 21st Century Demographic Data Needs—Implementing the American Community Survey: Report 3: Testing the Use of Voluntary Methods*. Washington, DC: U.S. Census Bureau.

U.S. Census Bureau, Geography Division (1999). *The Census Bureau's Master Address File (MAF): Census 2000 Address List Basics*. Washington, DC: U.S. Census Bureau.

U.S. Census Bureau, Geography Division (2000). *The Global Positioning Systems Test Project: A Report of the Newberry County, South Carolina Field Test*. Washington, DC: U.S. Census Bureau.

U.S. Census Bureau, Mobile Computing Device Working Group (2002). *Map Display Usability Test—Gloucester County, Virginia*. Washington, DC: U.S. Census Bureau.

U.S. Census Bureau, Public Information Office (2003). U.S. Census Bureau to count Americans in France, Kuwait and Mexico. Press release, January 16.

U.S. Department of Commerce, Office of Inspector General (2000). *Bureau of the Census: PAMS/ADAMS Should Provide Adequate Support for the Decennial Census, but Software Practices Need Improvement*. OSE-11684. Washington, DC: U.S. Department of Commerce.

U.S. Department of Commerce, Office of Inspector General (2002). *Improving Our Measure of America: What Census 2000 Can Teach Us in Planning for 2010*. OIG-14431. Washington, DC: U.S. Department of Commerce.

U.S. Department of Commerce, Office of Inspector General (2003). *Bureau of the Census: MAF/TIGER Redesign Project Needs Management Improvements to Meet Its Decennial Goals and Cost Objective*. OSE-15725. Washington, DC: U.S. Department of Commerce.

U.S. General Accounting Office (1997, February). *High Risk Series: Quick Reference Guide*. GAO/HR-97-2. Washington, DC: U.S. Government Printing Office.

U.S. General Accounting Office (2002a). *The American Community Survey: Accuracy and Timeliness Issues*. GAO-02-956R. Washington, DC: U.S. Government Printing Office.

U.S. General Accounting Office (2002b, April). *Legal Authority for American Community Survey*. Letter Report B-289852. Washington, DC: U.S. Government Printing Office.

Vitrano, F., J. Treat, and R. Pennington (2003, November 4). *Address List Development in Census 2000*. Census 2000 Testing, Experimentation, and Evaluation Program, Topic Report 8 (revised). Washington, DC: U.S. Census Bureau.

Waite, P. J. (2002). 2010 census reengineering. Presentation to the Panel on Research on Future Census Methods, September 13, 2002, Washington, DC. U.S. Census Bureau.

Whitworth, E. (2002, August 14). *Internet Data Collection*. Census 2000 Evaluation A.2.b. Washington, DC: U.S. Census Bureau.

Working Group on LUCA (2001). *Assessment of the 2000 Census LUCA Program*. Washington, DC: Committee on National Statistics.

Zaslavsky, A. M. and A. L. Schirm (1998). Interactions between the American Community Survey estimates and federal funding formulae: A look ahead. In *American Community Survey Workshop—Technical Papers*. Washington, DC: Committee on National Statistics, National Research Council.

Zaslavsky, A. M. and A. L. Schirm (2002). Interactions between survey estimates and federal funding formulas. *Journal of Official Statistics 18*(3), 371–392.

Biographical Sketches of Panel Members and Staff

Benjamin F. King *(Chair)* is a private statistical consultant, having retired in 1997 as professor of decision sciences at Florida Atlantic University. During his academic career he held tenured faculty positions in the Graduate School of Business of the University of Chicago and in both the School of Business Administration and the Department of Statistics at the University of Washington. In addition, he was director of survey methods at the Educational Testing Service. His research interests include survey sampling, Bayesian methods, and general applications of statistics to problems of business, public policy, and the law. A fellow of the American Statistical Association and elected member of the International Statistical Institute, he received his A.B., M.B.A., and Ph.D. from the University of Chicago. He has served on three previous panels of the Committee on National Statistics (CNSTAT).

David A. Binder is director general of the Methodology Branch at Statistics Canada, where he has held several positions since 1971. He is a past member of the Census Advisory Committee of Professional Associations (American Statistical Association subcommittee). His research interests include methods for treating nonresponse in surveys, variance estimation, innovation in government surveys, data analysis for complex surveys, and confidentiality of survey data. He is a fellow of the American Statistical Association and an elected member of the International Statistical Institute. He received his B.Sc. from the University of Toronto and his Ph.D. from Imperial College of Science and Technology.

Michael L. Cohen is a senior program officer for the Committee on National Statistics, currently serving as co-study director for the Panel on Research on Future Census Methods and staff for the Panel to Review the 2000 Census. He previously assisted the Panel on Estimates of Poverty for Small Geographic Areas and directed the Panel on Statistical Methods for Testing and Evaluating Defense Systems. Formerly, he was a mathematical statistician at the Energy Information Administration, an assistant professor in the School of Public Affairs at the University of Maryland, and a visiting lecturer in statistics at Princeton University. His general area of research is the use of statistics in public policy, with particular interest in census undercount, model validation, and robust estimation. A fellow of the American Statistical Association, he received a B.S. degree in mathematics from the University of Michigan and M.S. and Ph.D. degrees in statistics from Stanford University.

Daniel L. Cork is a program officer for the Committee on National Statistics, currently serving as co-study director of the Panel on Research on Future Census Methods and assisting the Panel to Review the 2000 Census. His research interests include quantitative criminology (particularly space-time dynamics in homicide), Bayesian statistics, and statistics in sports. He holds a B.S. degree in statistics from George Washington University and an M.S. in statistics and a joint Ph.D. in statistics and public policy from Carnegie Mellon University.

Mick P. Couper is a senior associate research scientist in the Survey Methodology Program in the Survey Research Center at the University of Michigan, an adjunct associate professor in the Department of Sociology at the University of Michigan, and a research associate professor in the Joint Program in Survey Methodology at the University of Maryland, College Park. He previously worked at the Census Bureau as a visiting researcher from 1992 to 1994. He has published in the areas of survey and census nonresponse and the use of computer technology for data collection. He received an M.Soc.Sc. from the University of Cape Town, an M.A. in Applied Social Research from the University of Michigan, and a Ph.D. in sociology from Rhodes University in Grahamstown, South Africa.

Shoreh Elhami *(consultant)* is the geographic information systems (GIS) director for the Delaware County, Ohio, Auditor's DALIS (Delaware Appraisal Land Information System) Project. Her involvement in census-related projects goes back to the 1990 census, when she became involved with the postcensus review process. From 1997 to 2000, as the county's census liaison, she orchestrated all Census 2000 related activities. She has more than 12 years of experience in the GIS field and is the principal architect of the DALIS Project's GIS, which was the primary resource used for the Local Update of Census Addresses (LUCA) in the county. Under her direction, the DALIS Project has received Ohio's Best GIS Practices Award from the Ohio Geographically Referenced Information Program in 1998 as well as the Environmental Systems Research Institute's Special Achievement Award in 2000. Elhami frequently speaks at national and international GIS conferences and is the past president of the Ohio Chapter of the Urban and Regional Information Systems Association. She served as a member of the LUCA Working Group jointly commissioned by the Panel to Review the 2000 Census and the Panel on Research on Future Census Methods. She is currently a member of the Mapping Science Committee of the National Research Council's (NRC) Division on Earth and Life Sciences. She received an M.A. in city and regional planning from Ohio State University.

C.A. "Al" Irvine is a private consultant in San Diego, California. He has experience in many areas of software engineering and information management dating back to the late 1950s. His most relevant experience is with System Development Corporation (1968–1970), NCR Corporation (1970–1973), as a co-founder of SofTech Microsystems (1979–1984), and as founder of Eclectic

Solutions (1984–present). He has consulted for IBM, Digital Equipment Corporation, Hewlett-Packard, Bank of America, ITT, Alcoa, NEC, Toshiba, and General Motors. He served as a panel member of the Computer Science and Telecommunications Board (CSTB) Committee to Review the Tax Systems Modernization of the Internal Revenue Service.

Sallie Keller-McNulty is group leader for the Statistical Sciences Group at Los Alamos National Laboratory. Prior to her move to Los Alamos, she was professor and director of graduate studies at the Department of Statistics, Kansas State University, where she had been on the faculty since 1985. She spent two years (1994–1996) as program director of Statistics and Probability, Division of Mathematical Sciences, National Science Foundation. She is an expert in the area of data access and confidentiality and currently serves on a National Research Council CSTB committee that is studying information technology and federal services. She is a fellow of the American Statistical Association and received the association's Founders' Award in 2002. She received her Ph.D. in statistics from Iowa State University.

George T. Ligler is a private consultant in Potomac, Maryland. He has extensive experience in information management and software and computer system engineering, as is evident from his work at Burroughs Corporation (1980–1982), Computer Sciences Corporation (1984–1988), and at GTL Associates, a private company that he founded. Most recently, he was a panel member of the CSTB Committee to Review the Tax Systems Modernization of the Internal Revenue Service. A Rhodes scholar, he received his B.S. in mathematics from Furman University in 1971 and his M.Sc. and D.Phil. from Oxford University.

Michael M. Meyer is chief scientist at Intelligent Results, Inc., a software company in Seattle, Washington. Previously, he held mathematics and engineering analyst positions at the Boeing Company and at Amazon.com. He has also served as senior research scientist in the Departments of Statistics and of Academic Computing and Media at Carnegie Mellon University, and held an academic appointment in the Department of Statistics at the University of Wisconsin at Madison. In 1991–1992, on a part-time basis, he served as the study director for the CNSTAT Panel to Review Evaluation Studies of Bilingual Education. His research interests include statistical computing and categorical data analysis. A fellow of the American Statistical Association, he received a B.A. (with honors) in mathematics from the University of Western Australia and a Ph.D. in statistics from the University of Minnesota.

Keith F. Rust is vice president and associate director at Westat, Inc. He is also a research associate professor in the Joint Program on Survey Methodology at the University of Maryland. He was formerly with the Australian Bureau of Statistics. He was a member of the Committee on National Statistics of the National Research Council (1992–1998). He has extensive experience in sampling methods, the design and specification of large-scale sample surveys, and analysis of survey data. He is a fellow of the American Statistical Association and a member of the International Statistical Institute. He served as a member of the NRC Panel to Evaluate Alternative Census Methodologies and as chair of the NRC Panel on Alternative Census Methods. He received a Ph.D. in biostatistics from the University of Michigan.

Joseph J. Salvo is director of the Population Division at the New York City Department of City Planning. He was previously deputy director and senior demographer. He worked at the U.S. Census Bureau in 1981–1982. He has broad experience in immigration, the application of small-area data for policies and programs, and the use of census data. A past president of the Association of Public Data Users, he has experience working with MAF/TIGER and the American Community Survey. Salvo chaired the Local Update of Census Addresses (LUCA) working group jointly sponsored by the Panel to Review the 2000 Census and the Panel on Research on Future Census Methods. He received his Ph.D. in sociology from Fordham University.

Joseph L. Schafer is associate professor of statistics at Pennsylvania State University. He was at the Census Bureau in the Statistical Support Division (1989–1991), and at the Bureau of

Labor Statistics on an American Statistical Association/National Science Foundation (ASA/NSF) research fellowship (1997). His main area of research involves the analysis of incomplete data. He received his Ph.D. in statistics from Harvard University.

Allen L. Schirm is a senior fellow and associate director at Mathematica Policy Research, Inc. His principal research interests include census methods, small-area estimation, sample and evaluation design, and the use of administrative records and survey data for policy analysis and program evaluation in the areas of food and nutrition and education and training policy. He has served on the CNSTAT Panel on Estimates of Poverty for Small Geographic Areas and the Panel on Formula Allocations. He received an A.B. in statistics from Princeton University and a Ph.D. in economics from the University of Pennsylvania.

Joseph H. Sedransk is professor of statistics at Case Western Reserve University. He has held previous tenured positions in the departments of statistics at Iowa State University, the University of Wisconsin, the State University of New York at Buffalo, the State University of New York at Albany, and the University of Iowa. He is a fellow of the American Statistical Association and an elected member of the International Statistical Institute. Relevant to census methods, he served as an ASA/NSF fellow at the Census Bureau, and as a member of the ASA advisory committee to the Census Bureau. He received his B.S. in economics from the University of Pennsylvania and his Ph.D. in statistics from Harvard University.

C. Matthew Snipp is a professor in the Department of Sociology at Stanford University. He has written extensively on American Indians [*American Indians: The First of the Land* (1989), *Research in Human Capital and Development: American Indian Economic Development* (1996)], and he has written specifically on the interaction of American Indians and the U.S. Census ["American Indians" in *Encyclopedia of the U.S. Census* (2000)]. He serves or has served on the Technical Advisory Committee on Racial and Ethnic Statistics and the Native American Population Advisory Committee, both of the U.S. Census Bureau. He also contributed a chapter on "The Size and Distribution of the American Indian Population: Fertility, Mortality, Residence, and Migration" in *Changing Numbers, Changing Needs: American Indian Demography and Public Health*, a 1996 publication of the National Research Council's Committee on Population. He received his M.S. and Ph.D. in sociology from the University of Wisconsin.

Donald Ylvisaker is an emeritus professor of statistics at the University of California, Los Angeles, having previously been on the faculties of Columbia University, New York University, and the University of Washington. His primary research interest is in the design of experiments; his applied interests have developed as a consulting statistician, frequently on legal matters. Relevant to census methods, he held a joint statistical agreement with the Census Bureau in 1990–1991 in which he reviewed the 1986 Test of Adjustment Related Operations (TARO); he served on the Advisory Panel to the Committee on Adjustment of Postcensal Estimates in 1992; and he was an associate editor of a special issue of the *Journal of the American Statistical Association* on census methods. A fellow of the American Statistical Association, he received his Ph.D. in statistics from Stanford University.

Alan M. Zaslavsky is a professor in the Department of Health Care Policy at Harvard Medical School. He has written extensively on issues concerning the decennial census, including weighting and administrative records. He is a fellow of the American Statistical Association. He has served on two Committee on National Statistics panels involving decennial census methodology—the Panel to Evaluate Alternative Census Methods (1992–1994) and the Panel on Alternative Census Methodologies (1995–1999)—as well as on the Panel on Estimates of Poverty for Small Geographic Areas (1996–2000). He also served on the Census Advisory Committee on Adjustment of Postcensal Estimates (1992). He received his M.S. degree in mathematics from Northeastern University and his Ph.D. in applied mathematics from the Massachusetts Institute of Technology.